the Rhubarb Cookbook

by
Jan Smart

first printing

Published and Distributed by
Gallimaufry Press
616 4th St.,
Nanaimo, B.C. V9R 1T7

printed by Hignell Printing Ltd.

cover photograph by Garry Smart

Copyright © Jan., 1991

ISBN 0-9694832-0-1

To all those who willingly offered up
their tastebuds time and time again along
the way and to you, Gentle Reader
Thank You
Jan

Pictured on the cover is Apple Rhubarb Pie

Rhubarb Cultivation

Rhubarb sprang from the same harsh demanding landscape that spawned Gengis Khan. It is very hard to kill and seems to thrive among the stones, broken tools and nettles at the end of the garden. An acquaintance (a non-rhubarb eater), told me about paving over the patch of rhubarb at the edge of her driveway when they widened it, only to have the rhubarb creep out from under the edge of the pavement the following spring.

Rhubarb must have a cold frosty dormant period in the winter in order to produce tender edible stalks. If it does not get this weather it will grow leggy, green and inedible.

Plant your rhubarb in your perennial bed. When it is grown in favourable conditions it will produce for years. Commercial growers like to turn their crop around every four years, but an aunt of mine enjoyed rhubarb from the same plant for fifteen years.

Rhubarb can grow quite large so it should be planted about 3 feet (1 meter) apart. Clear the space for the rhubarb plant of perennial weeds and dig a large deep hole. Line the hole with manure, mulch and topsoil. Put the root in and cover everything to a depth of one half an inch and tamp down the soil. Do not lime rhubarb, it prefers acidic soil. Water well and let nature take over.

Rhubarb does not grow well from seed. The usual method of propagation is root division. Find a friend or neighbour who is splitting a clump up or visit your local nursery. A word of caution here... find a type of rhubarb that will suit your family's needs. Rhubarb stalks come in sizes from the width of a woman's thumb to the size of a hefty walking stick. You may join the legions trying to send dinner guests home with a "little" rhubarb.

Rhubarb requires minimal care while growing. In winter, when it dies down, mulch it well with a layer of leaf-mould, compost or manure. We find that droppings from under the rabbit's cage works well. As long as you feed it like this it will give you years of enjoyment. Do not allow your rhubarb to flower; cut away the blossoms as the plant will divert energy from the leaves and stems to the production of unwanted blossoms.

In the spring move the mulch away from the crowns of the plant so the sun can warm the soil. If you would like to have rhubarb a little earlier in the spring, force the plants by covering them with buckets covered with fresh, long strawed manure. The heat from the manure decomposing will encourage growth. Uncover when the weather is warm enough for steady growth.

Harvesting Your Rhubarb

Use your rhubarb sparingly the first year of growth, to give it a good start. Pick the thick juicy stalks at the edges of the plant leaving the smaller stalks at the center to nourish the plant.

When you do pick the rhubarb, jerk the entire stalk away from the plant taking the pale bottom part as well. Do not cut or break off the stalks. This can let in rot.

Some people say not to harvest rhubarb after July. I have had rhubarb that was tender and tasty all summer long and rhubarb that grew thin and stringy after June. Let the behavior of your own plant be your guide. Do remember to leave enough growth on the clump to nourish the plant.

Rhubarb Pests

The rhubarb curculio is a coloured beetle about an inch (2.5 cm) long that bores into every part of the plant. When you find it, it can easily be picked off which avoids spraying with pesticides. This beetle also lives in dock plants so don't let dock plants grow anywhere near your rhubarb.

Rhubarb grows well in containers as long as they are large enough and have good drainage. It is important to compost your plant and to divide it when it seems to be outgrowing its pot. Divide rhubarb in early spring. Do not be shy when you divide the roots. We look for natural division in the crown and cut it apart with a shovel then move the severed piece to its own space.

We recently moved during winter and didn't bother taking a rhubarb root with us as there was rhubarb in the garden we were moving to. When the rhubarb came up in the spring it was a different type from what we were used to and we have since replaced it.

CONTENTS

Rhubarb Cultivation .... v–viii

Chapter #1. Sauces
1. Rhubarb Sauce 2
2. Rhubarb Pineapple Sauce 3
3. Rhubarb Orange Sauce 4
4. Rhubarb Ginger Sauce 5
5. Rhubarb Strawberry Sauce 6
6. Rhubarb Ham Glaze 7
7. Rhubarb Glaze for Poultry 8

Chapter #2. Family Favourites
8. Rhubarb Ginger Cake 10
9. Orange Rhubarb Cake 11
10. Rhubarb Pineapple Upside Down Cake 12
11. Rhubarb Bran Breakfast Cake 13
12. Honey Whole Wheat Rhubarb Cake 14
13. Quick Rhubarb Bundt Cake 15
14. Rhubarb Walnut Loaf 16
15. Rhubarb Loaf 17
16. Orange Rhubarb Loaf 18
17. Rhubarb Almond Muffins 19
18. Rhubarb Oatmeal Muffins 20
19. Tropical Rhubarb Oatmeal Bars 21
20. Rhubarb Oatmeal Bars 22
21. Rhubarb Meringue Squares 23
22. Rhubarb Cheesecake Squares 24
23. Rhubarb Coconut Bars 25
24. Rhubarb Raisin Squares 27
25. Rhubarb Strawberry Rolls 28

Chapter #3. Bettys, Crisps, and Cobblers
26. No Bake Cream Betty 30
27. Rhubarb Crisp Made with Cereal 31
28. Rhubarb Heavenly Hash 32
29. Rhubarb Cottage Pudding 33
30. Rhubarb Cake Pudding 34
31. Vicki's Rhubarb Orange Pudding 35
32. Rhubarb Tapioca Pudding 36
33. Sunshine Cobbler 37

34. Rhubarb Cobbler 38
35. Maple Rhubarb Cobbler 40
36. Rhubarb Crumble 41
37. Rhubarb Apple Crisp 42
38. Rhubarb Crisp 43
39. Rhubarb Crunch 44

Chapter #4. Company's Coming
40. Rhubarb On A Cloud 46
41. Strawberry Flan with Rhubarb Sauce 47
42. Rhubarb Walnut Cheesecake
    with Cointreau Sauce 49
43. Rhubarb Sponge Pudding 51
44. Strawberry Rhubarb Puff 52
45. Rhubarb Floating Islands 53
46. Rhubarb Chiffon Parfait 54
47. Rhubarb Bavarian Cream 55
48. Rhubarb Ginger Mousse 57
49. Rhubarb Strawberry Mousse 58
50. Rhubarb Cheesecake 59
51. Rhubarb Peach Compote 61
52. Honey Ginger Compote 62
53. Double Orange Rhubarb Compote 62
54. Rhubarb Strawberry Compote 63
55. Rhubarb Fool 64
56. Rhubarb Meringue Torte 65
57. Quick Rhubarb Cake in a Blanket 67

Chapter #5. Frozen Delights
58. Rhubarb Cointreau Sorbet 70
59. Frozen Rhubarb Cream
    with Strawberry Coulis 71
60. Rhubarb Yogurt-sicles 72
61. Rich Rhubarb Yogurt Creme 73
62. North South Ice Cream 74
63. Frozen Rhubarb Custard 75
64. Rhubarb Sherbet 76
65. Honey'n Rhubarb Ice Cream 77
66. Gingersnap Honey Ice Cream 78
67. Strawberry Rhubarb Ice Cream 79
68. Rhuberry Ice 80

69. Rhubarb Ice Cream 81
70. Rhubarb Cheesecake Ice Cream 82

Chapter #6. Jams, Jellies, and Preserves
71. Rhubarb Ginger Jam 84
72. Pineapple Rhubarb Jam 85
73. Raspberry Rhubarb Jam 87
74. Rhubarb Marmelade 88
75. Rhubarb Walnut Marmelade 89
76. Rhubarb Carrot Marmelade 90
77. Rhubarb Strawberry Conserve 91
78. Cinnamon Rhubarb Conserve 92
79. Spiced Rhubarb Jelly 93
80. Rhubarb Jelly 94
81. Rhuberry Jelly 95
82. Rhubarb Apricot Chutney 96
83. Rhubarb Raisin Chutney 97
84. Rhubarb Chutney
    made with Dates and Tomatoes 98
85. Rhubarb Fig Preserves 99
86. Rhubarb Onion Relish 100

Chapter #7. Pies and Tarts
87. Rhubarb Pie 102
88. Rhubarb Chiffon Pie 103
89. Rhubarb and Raisin Pie 104
90. Sour Cream Rhubarb Pie 105
91. Rhubarb Crumb Top Pie 106
92. Almost Rhubarb Pie 107
93. Rhubarb and Raisin Custard Pie 108
94. Apple Rhubarb Pie 109
95. Rhubarb Cream Pie 110
96. Rhubarb Strawberry Deep Dish Pie 111
97. Rhubarb Custard Pie 112
98. Rhubarb Crumb Tart 113
99. Rhubarb Cinnamon Tart 114
100. Rhubarb Butter Tarts 115
101. Rhubarb Strawberry Tarts 116
102. Rhubarb Custard Tarts 116

Chapter #8. Bits and Pieces

- 103. Scandinavian Style Rhubarb 118
- 104. Rhubarb Strawberry Tapioca 119
- 105. Rhubarb Barbequed Short Ribs 120
- 106. Chilled Rhubarb Soup 121
- 107. Rhubarb Fruit Leather 122
- 108. Harvard Style Rhubarb Beets 123
- 109. Rhubarb Cream Cheese Jellied Salad 124
- 110. Fluffy Lemon Orange Dressing 125
- 111. Jellied Rhubarb Strawberry Salad 126
- 112. Rhubarb Cocktail 127
- 113. Rhubarb Wine 128
- 114. Rhubarb Punch 129
- 115. Rhubarb Concentrate 130
- 116. Rhubarb Aphid Spray 131

Sauces

Rhubarb Sauce

1 pound	rhubarb, tops removed	500 mL
⅓ cup	sugar	75 mL
¼ tsp.	cinnamon	1.25 mL
⅛ tsp.	salt	.62 mL
1 tbsp.	water	15 mL

Trim off ends and any remaining leaves from rhubarb. Wash stalks carefully. Cut into 1″ pieces.

In a medium sized saucepan, combine rhubarb, sugar, cinnamon, water and salt. Heat to boiling. Stir frequently. Reduce heat and simmer until saucey — 15 to 20 minutes. Cool to room temperature and then refrigerate until ready to use.

* To thicken rhubarb sauce for use in recipes, mix 2 tbsps. of cornstarch with 2 tbsps. of cold water and stir into the stewed rhubarb mixture while simmering.

Yields: 1½ to 2 cups (375-500 mL).

Rhubarb Pineapple Sauce

2½ cups	rhubarb	625 mL
1 cup	sugar	250 mL
1 - 14 oz. can	crushed pineapple and juice	396 mL
2 tbsps.	quick cooking tapioca	30 mL

Trim off ends and any remaining leaves from rhubarb and wash stalks. Cut into 1 inch pieces.

In a medium saucepan, combine all of the sauce ingredients. Let stand for 5 minutes to let the tapioca soften.

Heat mixture to boiling over medium heat. Reduce heat and simmer until rhubarb is soft, about 25 minutes, stirring frequently.

Cool to room temperature. Pour into storage container and refrigerate until ready to use.

Yields: 4 cups (1 L)

Great on banana splits.

Rhubarb Orange Sauce

3 cups	rhubarb cut into 1" pieces	750 mL
½ cup	sugar	125 mL
⅓ cup	orange liqueur or orange juice	75 mL
1 tbsp.	coarsely grated orange rind	15 mL
¼ cup	water	60 mL

Combine all ingredients in a heavy saucepan. Heat slowly to boiling, stirring frequently to prevent scorching.

Reduce heat and simmer uncovered another 15 minutes, stirring occasionally. Add more water if sauce becomes too thick.

Yields: 2 cups (500 mL).

 Rhubarb is known to have been used by man since 2700 B.C.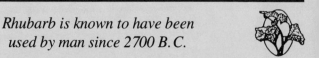

Rhubarb-Ginger Sauce

3 cups	rhubarb cut into ½" pieces	750 mL
⅓ cup	sugar	75 mL
⅓ cup	orange juice	75 mL
2 tbsps.	finely cut up candied ginger	30 mL

Combine rhubarb, sugar, orange juice and ginger in a large heavy saucepan. Heat to boiling stirring frequently to prevent scorching.

Reduce heat and simmer uncovered, stirring occasionally, until mixture is as thick as apple sauce, about 30 minutes.

If a stronger ginger taste is desired, add an additional ½ tsp. (2.5 mL) powdered ginger.

Yields: 1½ to 2 cups (375-500 mL).

Rhubarb Strawberry Sauce

4 cups	rhubarb cut into 1" pieces	1 L
¼ cup	sugar	60 mL
1 cup	fresh or frozen strawberries	250 mL
1 tbsp.	cornstarch	15 mL
½ tsp.	grated lemon rind	2.5 mL

Wash and cut up rhubarb. Wash and slice strawberries. Place all ingredients in large sized saucepan and cook over medium heat until rhubarb and strawberries are well blended, about 15 minutes, stirring occasionally to prevent scorching. Makes 2½ cups (625 mL).

Serve as a sauce over plain yogurt, ice cream, or white cake, or layer in parfait glasses with vanilla pudding for an easy, elegant dessert.

Rhubarb Ham Glaze

2 cups	rhubarb cut into ½″ (1 cm) pieces	500 mL
½ cup	brown sugar	125 mL
¼ cup	water	60 mL
½ tbsp.	powdered mustard	7.5 mL
¼ tsp.	cloves	1.25 mL

Wash and cut up rhubarb. Put rhubarb, water, sugar, mustard and cloves together in a small saucepan over medium heat. Simmer, stirring frequently until rhubarb mixture reaches a saucey consistency, about 20 minutes. Use on ham as you would a brown sugar or cranberry glaze.

Yields: 1 cup (250 mL).

Rhubarb Glaze for Poultry

¼ cup	water	60 mL
2 cups	rhubarb cut into ½″ (1 cm) pieces	500 mL
½ cup	sugar	125 mL
1 tbsp.	lemon juice	15 mL
1 tsp. or	fresh lemon thyme	5 mL
½ tsp.	dried lemon thyme	2.5 mL

Wash and cut up rhubarb. Combine rhubarb, water, sugar, lemon juice and lemon thyme together in a small saucepan over medium heat. Stir frequently to prevent scorching. Simmer until rhubarb is broken up and well blended, about 20 minutes.

Enjoy as a glaze over a whole roasting chicken or over oven-baked chicken parts.

Yields: 1 cup (250 mL).

Family Favourites

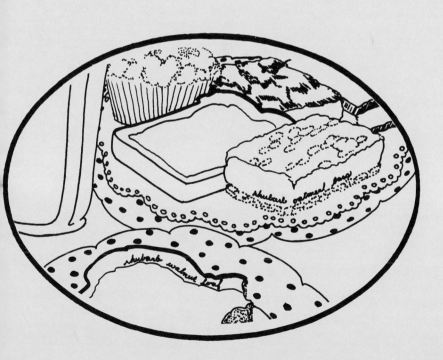

Rhubarb Ginger Cake   Oven 350°F

¼ cup	butter or margarine	60 mL
¼ cup	shortening	60 mL
1 ½ cups	sugar	375 mL
½ tsp	vanilla	2.5 mL
3	beaten eggs	3
2 cups	sifted all purpose flour	500 mL
½ tsp	ginger	2.5 mL
1 tsp	baking soda	5 mL
1 tsp	baking powder	5 mL
¼ tsp	salt	1.25 mL
½ cup	buttermilk	125 mL
1 cup	rhubarb ginger sauce (p. 5)	250 mL

Heat oven to 350°F. Grease a 9x13 inch (23x33 cm) baking pan.

Cream butter and shortening together. Gradually add sugar, creaming until light and fluffy.

Add eggs and vanilla, blending well.

Sift together flour, ginger, baking soda, baking powder, and salt.

Add dry ingredients to egg mixture alternately with rhubarb and buttermilk mixed together, beating well after each addition.

Pour batter into greased baking pan and bake at 350°F for 40 minutes or until center of cake springs back when gently pressed down with a finger.

Frost if desired. Serves 20.

Orange Rhubarb Cake Oven 350°F

2 cups	rhubarb cut into ½ " (1 cm) pieces	500 mL
3 cups	boiling water	750 mL
⅓ cup	orange marmalade	75 mL
½ cup	butter or margarine	125 mL
1 cup	sugar	250 mL
1	egg, beaten	1
1 tsp.	vanilla	5 mL
2 cups	all purpose flour	500 mL
1 tbsp.	baking powder	15 mL
1 cup	milk	250 mL
¼ tsp.	salt	1.25 mL

Wash and cut up rhubarb. Cover with boiling water and let stand for 5 minutes then drain. Stir marmalade into drained rhubarb. Set aside.

In a large bowl, cream together butter and sugar. Add the beaten egg and vanilla. Blend well.

Sift together flour, baking powder and salt. Grease a 9x9 inch (23x23 cm) baking pan.

Add sifted dry ingredients to creamed mixture alternately with milk.

Spread in greased baking pan and gently cover with rhubarb marmalade mixture.

Bake at 350°F for 45 to 50 minutes. Serves 16.

To vary use strawberry or raspberry jam instead of marmalade.

Rhubarb Pineapple Upside-Down Cake

Oven 350°F

2 tbsps.	butter or margarine	30 mL
½ cup	light brown sugar	125 mL
½ cup	pineapple chunks	125 mL
¾ cup	rhubarb cut into ½" (1 cm) pieces	185 mL
⅓ cup	shortening	75 mL
½ cup	sugar	125 mL
1	egg	1
1¼ cups	all purpose flour	310 mL
2 tsps.	baking powder	10 mL
½ tsp.	salt	2.5 mL
½ cup	pineapple syrup	125 mL
1 tsp.	vanilla	5 mL

Heat oven to 350°F. Grease an 8 × 8 inch (20.5 × 20.5 cm) baking pan.

Put butter and brown sugar in baking pan and place in oven until brown sugar is just melted.

Wash and chop rhubarb and put in pan along with pineapple chunks.

Cream together shortening, sugar, and egg.

Sift together flour, salt, and baking powder. Add to creamed mixture alternately with pineapple syrup. Add vanilla and mix well.

Spoon batter over rhubarb and pineapple. Bake at 350°F for 30 to 35 minutes. Remove from oven and let stand 10 minutes then invert onto serving plate. Serve warm with whipped cream or cool on its own. Serves 16.

To vary, used canned peach slices and syrup instead of pineapple chunks and syrup.

Rhubarb Bran Breakfast Cake   Oven 350°F

This is a lovely cake for a spring brunch.

⅓ cup	brown sugar	75 mL
½ cup	flaked almonds	125 mL
½ tsp	cinnamon	2.5 mL
½ cup	butter or margarine	125 mL
⅔ cup	brown sugar	150 mL
2	large eggs	2
1 tsp	vanilla	5 mL
1 cup	whole wheat flour	250 mL
½ cup	all purpose flour	125 mL
½ cup	bran	125 mL
1 tsp	baking powder	5 mL
1 tsp	baking soda	5 mL
1 cup	plain yogurt	250 mL
1 cup	diced rhubarb	250 mL

Heat oven to 350°F. Grease an 8x8 inch (20.5x20.5 cm.) baking pan.

Combine the first three ingredients in a small bowl and set aside.

Cream together brown sugar and butter. Add eggs one at a time, beating well after each addition. Add vanilla.

Sift together the whole wheat flour, all purpose flour, baking soda and baking powder. Stir in the bran mixing completely.

Add the dry mixture to the creamed mixture alternately with the yogurt.

Pour the batter into the greased baking pan. Top with the diced rhubarb.

Sprinkle this with the topping mixture of brown sugar, cinnamon and almonds.

Bake at 350°F for 35 to 40 minutes or until the rhubarb is soft and the cake underneath is done. Serves 9.

Honey Whole Wheat Rhubarb Cake

Oven 350°F

½ cup	softened butter or margarine	125 mL
½ cup	light brown sugar	125 mL
½ cup	honey	125 mL
1	egg	1
2 cups	unsifted cake flour	500 mL
½ cup	unsifted whole wheat flour	125 mL
½ tsp.	cinnamon	2.5 mL
½ tsp.	ginger	2.5 mL
2 tsps.	baking powder	10 mL
1 tsp.	baking soda	5 mL
¾ cup	buttermilk	185 mL
1 cup	rhubarb sauce (p. 2)	250 mL
	icing sugar for decoration	

Heat oven to 350°F. Grease an 8 inch springform pan. In a large bowl, cream butter, sugar, honey and egg together until well blended.

Measure and stir together cake flour, whole wheat flour, cinnamon, ginger, baking powder and baking soda.

Mix buttermilk and rhubarb sauce together.

Alternately add rhubarb and buttermilk and dry ingredients to creamed mixture. Pour into greased pan and bake for 40-50 minutes or until center of cake springs back when pressed with a fingertip.

Cook cake for 10 minutes before removing rim. Sprinkle with icing sugar when cool or ice as desired.

For a nice stencilled design lay a plastic or paper doily on top of cake. Sift icing sugar over doily and gently lift it off. Doily design will be stencilled on the top of the cake.

Quick Rhubarb Bundt Cake   Oven 350°F

1 - 18 oz. pkg.	vanilla cake mix	510 g.
7-10 drops	red food coloring	7-10 drops
1 cup	rhubarb sauce (p. 2)	250 mL

Heat oven to 350°F. Grease and flour bundt cake pan. Mix cake according to package directions.

Measure 1 cup (250 mL) cake batter into small bowl.

Add the rhubarb sauce and food coloring. Mix well.

Pour plain batter into bundt pan. Pour rhubarb batter on top of this distributing evenly around pan.

Gently run spatula down through the rhubarb mixture into the white batter to create a marbled effect.

Bake according to package directions or for 40 minutes at 350°F. (Until an indent pressed into the center of the cake disappears.)

Invert to remove the pan and ice when cool with your favourite vanilla icing. Serves 16.

Rhubarb Walnut Loaf

Oven 350°F

2 cups	sifted all purpose flour	500 mL
1 ½ tsps.	baking powder	7.5 mL
½ tsp.	salt	2.5 mL
½ tsp.	baking soda	2.5 mL
¾ cup	sugar	185 mL
¼ cup	finely chopped walnuts	60 mL
1	egg, well beaten	1
¾ cup	milk	185 mL
2 tbsps.	oil	30 mL
1 cup	rhubarb sauce (p. 2)	250 mL

Topping:

1 tbsp.	finely chopped walnuts	15 mL
½ tsp.	cinnamon	2.5 mL
3 tbsps.	light brown sugar	45 mL

Heat oven to 350°F. Grease a 5×3×9.5 inch (13×7.5×24 cm) loaf pan.

Sift together all purpose flour, baking soda, salt and baking powder.

Dredge walnuts with flour mixture.

Stir in sugar.

Mix milk, beaten egg and oil together. Stir into dry ingredients until just moistened.

Add rhubarb sauce and stir into dough until just evenly mixed. Pour into loaf pan.

Mix together topping and sprinkle over dough.

Bake at 350°F for 50 minutes or until done. Let stand overnight to mellow flavour. Makes 1 loaf.

Rhubarb Loaf Oven 350°F

1½ cups	packed brown sugar	375 mL
1 cup	oil	250 mL
1	egg, beaten	1
1 cup	buttermilk	250 mL
3 cups	all purpose flour	750 mL
1 tsp.	baking soda	5 mL
1 tsp.	salt	5 mL
2 tsps	cinnamon	10 mL
1½ cups	rhubarb cut into ½" (1.25 cm) pieces	375 mL
½ cup	granulated sugar	125 mL
1 tbsp.	melted butter	15 mL

Mix together oil and brown sugar.

Beat in egg and buttermilk.

Sift together flour, salt, soda and 1 tsp. (5 mL) cinnamon.

Stir into creamed mixture. Fold in rhubarb.

Spoon batter into two small greased loaf pans.

Mix together granulated sugar, melted butter and remaining cinnamon. Sprinkle over loaves.

Bake at 350°F for 1 hour. Makes 2 small loaves.

Orange Rhubarb Loaf

Oven 350°F

2 cups	all purpose flour	500 mL
¾ cup	sugar	185 mL
1½ tsp.	baking powder	7.5 mL
½ tsp.	baking soda	2.5 mL
1 tsp.	salt	5 mL
1	egg	1
1 cup	rhubarb sauce (p. 2)	250 mL
¾ cup	orange juice	185 mL
2 tbsps.	oil	30 mL

Heat oven to 350°F. Grease a 9.5×5×3 inch (24×13×8 cm) loaf pan.

Sift together flour, sugar, baking powder, baking soda and salt.

Stir the rhubarb sauce (cool), orange juice and oil.

Beat the egg with a fork and stir into the liquid mixture. Blend well.

Add dry ingredients to wet ingredients stirring just 'til moistened.

Bake in loaf pan at 350°F for 50 minutes. Remove from pan and cool.

Icing —

¾ cup	icing sugar	185 mL
1 tbsp.	orange juice	15 mL
½ tsp.	grated orange peel	2.5 mL

Blend icing sugar, orange juice and grated orange peel in blender until smooth.

Drizzle over loaf. Serves 12.

Rhubarb Almond Muffins   Oven 350°F

This is my family's favourite muffin recipe

1 cup	rhubarb sauce (p. 2)	250 mL
2½ cups	unsifted all purpose flour	625 mL
½ cup	slivered almonds	125 mL
1 tsp.	baking powder	5 mL
1 tsp.	baking soda	5 mL
½ tsp.	salt	2.5 mL
¼ tsp.	nutmeg	1.25 mL
¼ tsp.	cinnamon	1.25 mL
¾ cup	light brown sugar	185 mL
1 cup	buttermilk	250 mL
½ cup	oil	125 mL
1	large egg	1

Heat oven to 375°F. Grease 12 muffin cups.

In a large bowl, mix flour, almonds, baking powder, baking soda, salt, nutmeg and cinnamon.

In a small bowl, beat brown sugar, buttermilk, oil and egg together until well mixed. Add rhubarb sauce and stir well.

Add all at once to dry ingredients. Stir quickly only until dry ingredients are moistened.

Spoon batter into greased muffin cups. Fill cups ⅔ full and bake at 375°F for 18 to 20 minutes.

Yields: 12 muffins.

Rhubarb Oatmeal Muffins — Oven 350°F

½ cup	oil	125 mL
¾ cup	light brown sugar	185 mL
1	egg	1
1 cup	all purpose flour	250 mL
1 tsp.	baking powder	5 mL
1 tsp.	baking soda	5 mL
¼ tsp.	salt	1.25 mL
1 cup	rolled oats	250 mL
¾ cup	rhubarb sauce (p. 2)	185 mL

Beat together oil, egg and sugar. Stir in rhubarb sauce.

Sift together the dry ingredients. Add to creamed mixture.

Stir in rolled oats.

Fill greased muffin tins ¾ full.

Bake in a 350°F oven for 25-30 minutes.

Yields: 12 muffins.

On July 17, 1947 the U.S. Customs Court of Buffalo, New York declared that rhubarb, though technically a vegetable, would be considered a fruit for the purpose of determining the level of import duty collected

Tropical Rhubarb Oatmeal Bars

Oven 375°F

½ cup	butter or margarine	125 mL
1 cup	brown sugar	250 mL
1½ cups	sifted all purpose flour	375 mL
½ tsp.	baking soda	2.5 mL
1½ cups	quick cooking rolled oats	375 mL
1 tbsp.	water	15 mL
1 tbsp.	grated orange peel	15 mL
½ cup	grated coconut	125 mL
3 cups	thickened rhubarb pineapple sauce (p. 3)	375 mL
1 tsp.	salt	5 mL

Heat oven to 375°F. Lightly grease a 13×9 inch (33×23cm) baking pan.

In a medium sized bowl, combine flour, rolled oats, brown sugar, orange peel, salt, baking powder, water and coconut.

Add softened butter and work in until the mixture forms coarse crumbs.

Press half of the crumb mixture into the bottom of the greased pan.

Spread with the thickened rhubarb pineapple sauce.

Top with the remaining crumb mixture.

Bake 25 to 30 minutes or until the top is golden brown and the fruit is bubbly. Cool to room temperature before cutting. Serves 20.

Rhubarb Oatmeal Bars   Oven 375°F

½ cup	butter or margarine	125 mL
1 cup	brown sugar	250 mL
1½ cups	sifted all purpose flour	375 mL
1 tsp.	salt	5 mL
½ tsp.	baking soda	2.5 mL
1½ cups	quick cooking rolled oats	375 mL
1 tbsp.	water	15 mL
2 tsps.	cinnamon	10 mL
½ cup	flaked almonds	125 mL
3 cups	thickened rhubarb sauce (p. 2)	750 mL

Heat oven to 375°F. Lightly grease a 9×13 inch (33×23 cm) baking pan.

In a medium sized bowl, combine flour, rolled oats, brown sugar, cinnamon, salt, baking soda, and flaked almonds.

Add softened butter and water. Work in until the mixture forms coarse crumbs.

Press half of the crumb mixture into the bottom of the greased pan.

Spread with the thickened rhubarb sauce. Top with the remaining crumb mixture.

Bake 25 to 30 minutes or until the top is golden brown and the fruit is bubbly. Cool to room temp. before cutting. Serves 20.

Rhubarb Meringue Squares Oven 350°F

Crust:

¼ cup	butter or margarine	60 mL
¼ cup	sugar	60 mL
1	egg	1
1 tsp.	vanilla	5 mL
1¼ cup	flour	310 mL
1 tsp.	baking powder	5 mL
¼ tsp.	salt	1.25 mL

Filling:

2 cups	rhubarb cut into 1″ (2.5 cm) pieces	500 mL
2 tbsps.	water	30 mL
⅓ cup	sugar	75 mL
¼ tsp.	cinnamon	1.25 mL
2 tbsps.	cornstarch	30 mL
1 tbsp.	cold water	15 mL

Meringue Topping:

2	egg whites	2
½ cup	sugar	125 mL

Heat oven to 350°F. Grease a 9×9 inch (23×23 cm) baking pan.

In a medium sized bowl, cream together butter and egg. Add sugar and vanilla. Sift together flour, baking powder and salt. Mix into creamed mixture on low speed until a thick dough is formed. Pat into greased baking pan extending dough ½ inch (1 cm) up sides of pan. Bake at 350°F for 15 minutes.

Wash and chop rhubarb. In a small saucepan, combine rhubarb, sugar and 2 tbsps. water and cook over low to medium heat until rhubarb is saucey. Mix cornstarch and cold water together and add to sauce mixture. Cook until sauce is clear and thickened. Spread over baked crust.

Whip egg whites in a small bowl until soft peaks form. Beat in sugar. Spread on top of rhubarb sauce and bake at 350°F for 10 minutes or until meringue is golden brown. Cool before serving. Serves 16.

Rhubarb Cheesecake Squares — Oven 350°F

Crust:

1 cup	graham wafer crumbs	250 mL
2 tbsps.	sugar	30 mL
¼ cup	butter or margarine	60 mL
¼ tsp.	nutmeg	1.25 mL

Filling:

1 - 8 oz. pkg.	cream cheese	1 - 250 g
½ cup	sugar	125 mL
2 tbsps.	flour	30 mL
1	egg	1
⅔ cup	whipping cream	150 mL
2 tbsps.	lemon juice	30 mL

Topping:

2 cups	rhubarb cut into 1" (2.5 cm) pieces	500 mL
¾ cup	sugar	185 mL
2 tbsps.	water	30 mL
1 tbsp.	cornstarch	15 mL
1 tbsp.	cold water	15 mL

Wash and cut up rhubarb. Put rhubarb, sugar and 2 tbsps. (30 mL) water in a small saucepan over low to medium heat and cook until rhubarb is saucey. Mix cornstarch and cold water together and add to rhubarb sauce. Cook, stirring frequently, until sauce clears and thickens. Remove from heat and cool completely.

Heat oven to 350°F and grease an 8×8 inch (20.5×20.5 cm) baking pan. Combine graham wafer crumbs, sugar, nutmeg and butter in a small bowl. Press crumb mixture into bottom of pan.

Cream cheese until light and fluffy. Beat in sugar, egg and flour. Gradually beat in whipping cream. Add lemon juice. Pour onto prepared crust. Bake at 350°F for 35 to 40 minutes or until golden brown. Cool completely and then spread with rhubarb sauce. Chill well before serving. Serves 12 or 16.

Rhubarb Coconut Bars   Oven 350°F

Crust:
1 cup	all purpose flour	250 mL
½ cup	butter or margarine	125 mL
1 tsp.	baking powder	5 mL
1	egg	1
1 tsp.	milk	5 mL

Filling:
2 cups	rhubarb cut into 1″ (2.5 cm) pieces	500 mL
2 tbsps.	water	30 mL
⅓ cup	sugar	75 mL
1 tbsp.	cornstarch	15 mL
2 tbsps.	cold water	30 mL

Topping:
1 cup	sugar	250 mL
¼ cup	butter or margarine	60 mL
1 tsp.	vanilla	5 mL
1 egg		1
2 cups	grated coconut	500 mL

Filling:
Wash and cut up rhubarb. Combine rhubarb, sugar, and 2 tbsps. (30 mL) water in a small saucepan. Cook over medium heat until rhubarb is saucey, stirring occasionally to prevent scorching. When rhubarb

is cooked, mix together cornstarch and cold water. Add to rhubarb and cook until sauce clears and thickens. Set aside to cool.

Heat oven to 350°F. Grease an 8×8 inch (20.5×20.5 cm) baking pan.

Crust:
Sift together flour and baking powder in a medium sized bowl. Cut in butter. Beat egg and milk together. Stir into flour mixture until well blended. Pat into bottom of greased baking pan.

Pour cooled rhubarb sauce over crust.

Topping:
For the topping, cream together sugar and butter. Beat in vanilla and the egg. Stir in the grated coconut.

Spread over the rhubarb sauce.

Bake at 350°F for 30 minutes. Cool to room temperature before cutting. Serves 16.

 Arab and Jewish traders carried rhubarb and cinnamon from China into ancient Baghdad loaded on camel caravans.

Rhubarb Raisin Squares Oven 375°F

1	recipe pie crust for double crust pie	1
2 cups	rhubarb, cut into 1″ (2.5 cm) pieces	500 mL
1 cup	raisins	250 mL
1 cup	sugar	250 mL
1	large egg	1
2 tbsps.	cornstarch	30 mL
1 tbsp.	grated lemon rind	15 mL

Heat oven to 375°F. Grease a 9 inch (23 cm.) square baking pan.

Prepare pie crust. Divide in half. Roll out one half on a floured surface to make a 9 inch (23 cm) square. Fit this into the bottom of the greased pan.

In a medium sized bowl, combine rhubarb, raisins, sugar, egg, cornstarch, and lemon rind. Stir together well and spoon mixture over pastry in the pan.

Roll out remaining pie dough into a 9 inch (23 cm) square. Place over rhubarb mixture. Gently score top crust into 16 squares with a knife. Cut a small slit in the center of each square.

Bake at 375°F for 55 minutes or until pastry is golden brown and filling is bubbly. Cool to room temperature before cutting. Serves 16.

Rhubarb Strawberry Rolls   Oven 450°F

¾ cup	sugar	185 mL
1 cup	water	250 mL
2 cups	all purpose flour	500 mL
1 tbsp.	sugar	15 mL
3 tsps.	baking powder	15 mL
½ tsp.	salt	2.5 mL
⅓ cup	butter or margarine	75 mL
1	egg, beaten	1
⅔ cup	milk	150 mL
1 cup	sliced fresh strawberries	250 mL
2 cups	diced rhubarb	500 mL
⅓ cup	sugar	75 mL
1 tbsp.	melted butter or margarine	15 mL

Heat oven to 450°F.

Add ¾ cup sugar to 1 cup water. Simmer in a small saucepan for 5 minutes. Pour into a 9×9×2 inch (23×23×5 cm) baking pan.

Sift together flour, 1 tbsp. sugar, baking powder and salt. Cut in butter until the mixture forms coarse crumbs.

Combine milk and beaten egg. Add all at once to the dry ingredients, stirring only to moisten. Turn out onto a lightly floured surface and knead gently for ½ minute.

Roll out into a rectangle 13×8 inches (33×20.5 cm). Brush with 1 tbsp. melted butter. Top with strawberries and rhubarb. Sprinkle with ⅓ cup sugar.

Roll up like a jelly roll. Cut into 12 slices and arrange these on top of syrup in baking pan.

Bake at 450°F for 25 to 30 minutes. Delicious served warm with ice cream. Serves 12.

Bettys Crisps and Cobblers

No-Bake Cream Betty

2 cups	rhubarb, cut into 1″ (2.5 cm) pieces	500 mL
½ cup	sugar	125 mL
	water if needed	
¼ cup	butter	60 mL
3 cups	soft fresh white bread crumbs	750 mL
¼ cup	brown sugar	60 mL
⅔ cup	whipping cream	150 mL
1 - 3 oz pkg.	instant vanilla pudding	85 g

Combine rhubarb and sugar (water if needed) in a saucepan and cook over low heat until rhubarb is soft but unbroken. Cool.

Heat butter in a frying pan and toss the bread crumbs in this until very crisp and golden. Cool, then mix with sugar. Prepare pudding according to package directions.

Fill a dish with alternate layers of cool well-drained fruit and pudding. Cover with ¾ crumb mixture.

Whip cream lightly and spread over top, then decorate the top with a ring of crumbs. Chill and serve. Serves 4-6.

One pound of fresh rhubarb makes about two cups of cooked rhubarb.

Rhubarb Crisp
Made with Cereal

Oven 350°F

A rhubarb custard base topped with a crunchy cereal topping.

6 cups	rhubarb cut into 1″ (2.5 cm) pieces	1500 mL
4 cups	boiling water	1000 mL
2	eggs	2
¼ cup	flour	60 mL
½ tsp.	vanilla	2.5 mL
2 cups	dry bread crumbs	500 mL
1 cup	rice crispies	250 mL
½ cup	slightly crushed cornflakes	125 mL
½ cup	cheerios	125 mL
1 tsp.	nutmeg	5 mL
½ cup	brown sugar	125 mL
½ cup	melted margarine or butter	125 mL

Heat oven to 350°F. Grease a 9×13 inch (23×33 cm) baking pan. Wash and chop rhubarb. In a large oven-proof bowl or pot, cover rhubarb with boiling water. Let stand 5 minutes. Drain.

Meanwhile, blend eggs, sugar, vanilla and flour together until creamy. Stir in drained, slightly cooled rhubarb. Pour into baking dish.

In a large bowl, stir together bread crumbs, rice crispies, corn flakes, cheerios, nutmeg and brown sugar. Pour melted margarine over crumb mixture and stir until evenly mixed. Sprinkle over rhubarb.

Bake at 350°F for 35-45 minutes. Serves 12-15.

Rhubarb Heavenly Hash

3 cups	rhubarb	750 mL
½ cup	sugar	125 mL
¼ tsp.	cinnamon	1.25 mL
⅛ tsp.	salt	.62 mL
¼ tsp.	water	1.25 mL
2 tbsps.	cold water	30 mL
2 tbsps.	corn starch	30 mL
1 cup	whipping cream	250 mL
2 tbsps.	sugar	30 mL
½ tsp.	vanilla	2.5 mL
½ cup	miniature marshmallows	125 mL
¼ cup	sweetened shredded coconut	60 mL
¼ cup	toasted flaked almonds	60 mL

In a medium saucepan, combine rhubarb, sugar, cinnamon, salt and water. Heat to boiling stirring frequently. Reduce heat and simmer for 5 to 8 minutes or until fruit is saucey. Mix cornstarch and water. Stir into rhubarb sauce and cook until mixture is clear and thickened. Stir constantly. Chill.

In a medium sized bowl, whip cream, sugar and vanilla until stiff peaks form. Fold in marshmallows and coconut.

In a serving dish, layer one quarter of the cream, one third of the rhubarb sauce and one quarter of the almonds. Repeat twice. Top with remaining cream and almonds. Refrigerate until ready to serve.

Yields: 4-6 servings.

Rhubarb Cottage Pudding Oven 350°F

1 cup	fresh rhubarb cut into ½" (1 cm) pieces	250 mL
1 cup	sugar	250 mL
¼ cup	butter or margarine	60 mL
1	large egg	1
¼ tsp.	grated lemon rind	1.25 mL
1¾ cups + 2 tbsps.	unsifted cake flour	365 mL
2 tsps.	baking powder	10 mL
¼ tsp.	salt	1.25 mL
1 cup	milk	250 mL

Heat oven to 350°F. Grease an 8×8 inch baking pan. In a medium sized bowl, beat sugar and butter until fluffy. Beat in egg and lemon rind.

In a small bowl, mix flour, baking powder and salt. Add to sugar mixture alternately with milk. Beat well after each addition.

Dredge rhubarb in flour.

Fold in rhubarb pieces. Spoon into greased pan.

Bake 40 to 45 minutes or until center springs back when gently pressed with a fingertip.

Cut into 6 squares and serve with rhubarb sauce and whipped cream or ice cream.

Rhubarb Cake Pudding   Oven 350°F

Makes its own caramel sauce!

⅓ cup	butter	75 mL
1 cup	brown sugar	250 mL
4 cups	rhubarb cut into ½″ (1 cm) pieces	1000 mL
1 tbsp.	grated orange peel	15 mL
1⅓ cups	sifted all purpose flour	325 mL
1 cup	sugar	250 mL
2 tsps.	baking powder	10 mL
½ tsp.	salt	2.5 mL
⅓ cup	soft shortening	75 mL
⅔ cup	milk	150 mL
1 tsp.	vanilla	5 mL
½ tsp.	lemon extract	2.5 mL
1	egg	1

Heat oven to 350°F.

Put butter in 2 quart baking dish (2 L). Set in oven until butter is melted.

Sprinkle melted butter with brown sugar and rhubarb. Sprinkle grated orange peel over all.

Sift flour, sugar, baking powder and salt together into a bowl. Add shortening, milk, vanilla and lemon extract and beat for 2 minutes at medium speed. Add egg and beat well. Pour batter over fruit.

Bake about 40-45 minutes or until the cake springs back when touched lightly in the center. Spoon into serving dishes and serve hot, topped with whipped cream. Serves 6-9.

Vicki's Rhubarb Orange Pudding   Oven 350°F

⅔ cup	butter or margarine	150 mL
1 cup	brown sugar	250 mL
3 cups	cut up rhubarb	750 mL
1 tbsp.	coarsely grated orange peel	15 mL
¼ cup	flaked almonds	60 mL
1⅓ cups	all purpose flour	325 mL
1 cup	sugar	250 mL
2 tsps.	baking powder	10 mL
½ tsp.	salt	2.5 mL
⅓ cup	soft shortening	75 mL
⅔ cup	milk	150 mL
1 tsp.	vanilla	5 mL
½ tsp.	orange extract	2.5 mL

Heat oven to 350°F. Put butter in a 9×9×2 inch (23×23×5 cm) cake pan. Set in oven until butter is melted.

Sprinkle melted butter with brown sugar, rhubarb, almond flakes and grated orange peel.

Sift flour, sugar, baking powder and salt together into a bowl. Cut in shortening finely. Add milk, vanilla and orange extract. Mix at medium speed for 2 minutes.

Pour batter over fruit. Bake about 40 mins. or until batter springs back when touched lightly with a finger.

Serve with whipped cream if desired. Serves 6-9.

Rhubarb Tapioca Pudding

3 cups	rhubarb cut into ½ " (1 cm) pieces	750 mL
½ cup	quick cooking tapioca	125 mL
1¾ cups	water	435 mL
¼ cup	cointreau or orange juice	60 mL
¼ tsp.	salt	1.25 mL
1¾ cup	sugar	435 mL
2 tsps.	grated orange rind	10 mL

Combine water, cointreau, salt and tapioca in a medium saucepan. Slowly bring to a boil over low heat. Stir frequently. When thickened, add rhubarb, sugar and orange rind. Cook until rhubarb is saucey and tapioca is translucent.

Place in serving bowl and cool in fridge until set.

Very nice topped with sliced bananas and whipped cream. Serves 6-8.

Sunshine Cobbler

Oven 450°F

1½ cups	rhubarb cut into ½" (1 cm) pieces	375 mL
⅓ cup	water	75 mL
1 cup	canned crushed pineapple and juice	250 mL
½ cup	sugar	125 mL
½ cup	frozen orange juice concentrate	125 mL
2 tsps.	grated grapefruit rind	10 mL

Topping:

1¼ cups	all purpose flour	310 mL
2½ tsps.	baking powder	12.5 mL
¼ tsp.	salt	1.25 mL
3 tbsps.	brown sugar	45 mL
¼ cup	butter or margarine	60 mL
5 tbsps.	sour cream	75 mL

Heat oven to 450°F. Grease a 9×9 inch (23×23 cm) baking pan. In a medium sized saucepan, simmer water and rhubarb until tender. Add crushed pineapple and sugar and bring to a boil. Remove from heat and set aside.

Combine frozen orange juice concentrate and grated grapefruit rind. Spread this over the bottom of the greased baking pan. Add the hot fruit mixture.

Mix together the flour, baking powder, salt and brown sugar. Cut in butter or margarine until the mixture resembles coarse crumbs. Stir in sour cream.

Stir with a fork to form a soft dough. Turn onto a lightly floured surface and knead gently 8-10 times.

Cut into 6 pieces and place on top of the fruit mixture.

Bake at 450°F oven for 15-18 minutes.

Serve warm with whipped cream or ice cream. Serves 6.

Rhubarb Cobbler

Oven 400°F

Topping:

1 cup	sifted all purpose flour	250 mL
2 tbsps.	granulated sugar	30 mL
1½ tsps.	baking powder	7 mL
¼ tsp.	salt	1.25 mL
¼ cup	butter or margarine	60 mL
1	egg, lightly beaten	1
¼ cup	milk	60 mL

Filling:

1½ cups	granulated sugar	375 mL
2 tbsps.	cornstarch	30 mL
4 cups	rhubarb cut into 1″ (2.5 cm) pieces	1 L
2 tbsps.	water	30 mL
1 tbsp.	butter or margarine	15 mL
1 tbsp.	sugar	15 mL
¼ tsp.	cinnamon	1.25 mL

Heat oven to 400°F. Grease a 9 × 5 inch (23 × 13 cm) loaf pan or baking dish.

Combine sugar and cornstarch in a saucepan with water and rhubarb. Cook over low heat until rhubarb is tender and cornstarch is thickening. Remove from heat and stir in butter.

Heat oven to 400°F. Grease a 9 × 5 inch (23 × 13 cm) loaf pan or baking dish.

Combine sugar and cornstarch in a saucepan with water and rhubarb. Cook over low heat until rhubarb is tender and cornstarch is thickening. Remove from heat and stir in butter.

Sift together flour, sugar, salt and baking powder. Roughly cut in ¼ cup butter or margarine. Mix egg and milk together in a bowl and add to flour mixture. Stir to moisten, but do not to beat.

Pour cooked rhubarb into the baking dish and top with 6 tablespoons of batter.

Mix together remaining tablespoon of sugar and cinnamon. Sprinkle over batter.

Bake for 25 minutes.

Serve warm topped with ice cream, whipped cream or plain yogurt. Serves 4-6.

 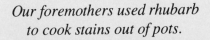

Our foremothers used rhubarb to cook stains out of pots.

Maple Rhubarb Cobbler

Oven 375°F

4 cups	rhubarb cut into ½″ (1 cm) pieces	1 L
1 - 14 oz. can	apricot halves, drained	378 mL
1 cup	pure maple syrup*	250 mL
1	egg	1
¾ cup	milk	185 mL
1 tbsp.	sugar	5 mL
2 cups	all purpose flour	500 mL
1 tbsp.	baking powder	15 mL
½ tsp.	salt	2.5 mL
⅓ cup	shortening	75 mL

Heat oven to 375°F. Grease an 8 cup (2 L) baking dish. Wash and chop rhubarb. Cover rhubarb with boiling water and let stand 5 minutes. Drain. Put apricot halves, rhubarb and maple syrup in baking dish. Set aside.

Whisk together egg and milk in a small bowl. Set aside.

In a medium sized bowl, mix together flour, sugar, salt and baking powder. Cut in shortening until mixture is crumbly. Make a well in the center of the dry ingredients. Pour in milk and egg and stir with a large spoon until just mixed. Do not over beat. Drop large spoonfuls of batter over fruit mixture in baking dish. Leave space between spoonfuls and do not spread batter evenly.

Bake at 375°F for 40-45 minutes or until top is golden brown. Serve warm with whipped cream or vanilla ice cream.

* If unavailable, substitute 1 cup of pancake syrup and 1 tsp. maple flavouring.

To vary, use peaches instead of apricots.

Rhubarb Crumble

Oven 375°F

3 cups	rhubarb cut into 1″ (2.5 cm) pieces	750 mL
2 tbsps.	orange juice	30 mL
1 tsp.	grated orange rind	5 mL
¾ cup	sugar	185 mL
⅓ cup	butter or margarine	75 mL
⅓ cup	brown sugar	75 mL
⅔ cup	flour	150 mL
⅛ tsp.	salt	.62 mL
¼ tsp.	baking soda	1.25 mL
⅔ cup	quick cooking oatmeal	150 mL

Heat oven to 375°F. Grease an 8×8×2 inch (20.5×20.5×5 cm) baking pan.

Wash and chop rhubarb. Put in bottom of baking pan. Sprinkle with orange juice and grated orange rind, sugar and cinnamon.

Dot with 1 tbsp. (15 mL) butter.

For Topping:
Combine brown sugar, flour, salt and baking soda. Cut in remaining butter. Stir in quick cooking oatmeal until evenly mixed. Spread over rhubarb.

Bake at 375°F for 40 minutes. Serves 6.

Rhubarb Apple Crisp Oven 350°F

3 cups	rhubarb cut into 1″ (2.5 cm) pieces	750 mL
3	large apples	3
½ cup	raisins	125 mL
1 cup	white sugar	250 mL
1 cup	brown sugar	250 mL
1 cup	oatmeal	250 mL
¾ cup	flour	185 mL
1 tsp.	cinnamon	5 mL
¼ tsp.	salt	1.25 mL
⅓ cup	margarine or butter	75 mL

Heat oven to 350°F. Grease an 8 cup (2 L) casserole dish. Wash and chop rhubarb. Peel and slice apples. In a large bowl, mix apples, rhubarb, raisins and white sugar together. Pour into greased casserole.

For Topping:
In a medium sized bowl, mix together brown sugar, flour, oatmeal, cinnamon and salt. Cut in margarine until mixture is well blended.

Pour over fruit. Bake in a 350°F oven for 40 minutes or until apples are tender. Serves 6-8.

Rhubarb is also known as "Lemon of the North".

Rhubarb Crisp

Oven 350°F

5 cups	rhubarb	1250 mL
1 cup	brown sugar	250 mL
1 tsp.	lemon juice	5 mL
⅔ cup	flour	150 mL
¾ cup	sugar	185 mL
¼ tsp.	salt	1.25 mL
¼ tsp.	allspice	1.25 mL
⅓ cup	margarine or butter	75 mL

Heat oven to 350°F. Grease an 8 × 8 × 2 inch (20.5 × 20.5 × 5 cm) baking pan.

Wash and chop rhubarb into 1 inch (2.5 cm) pieces.

Stir together rhubarb, brown sugar and lemon juice. Pour into greased baking pan.

For Topping:
Mix together flour, sugar, salt and allspice. Cut in butter until mixture is crumbly.

Sprinkle over fruit. Bake at 350°F for 40 minutes or until top is brown and crisp. Serves 6.

Rhubarb Crunch Oven 350°F

Served with vanilla ice cream, this is my Dad's favourite.

½ cup	raisins	125 mL
1 cup	boiling water	250 mL
3½ cups	rhubarb	875 mL
¾ cup	brown sugar	185 mL
½ tsp.	cinnamon	2.5 mL
2 tbsps.	flour	30 mL

Topping:

1 tbsp.	butter	15 mL
1 cup	sugar	250 mL
1 cup	flour	250 mL
1 tsp.	baking powder	5 mL
1	egg	1

Heat oven to 350°F. Grease an 8×8 inch (20.5×20.5 cm) baking pan. Pour boiling water over raisins and let stand for 5 minutes. Wash and chop rhubarb. Put in a large bowl. Drain raisins. Add to rhubarb. Mix in brown sugar, cinnamon and flour.

For Topping:
Dot with butter. Mix together flour, sugar and baking powder. Beat the egg and pour into rest of topping. Mix in until well blended. Pour on top of fruit. Bake at 350°F for 40 minutes. Serves 6-8.

Company's Coming

Rhubarb on a Cloud Oven 350°F

This truly decadent dessert can be made ahead in easy stages.

6	large egg whites	6
½ tsp.	cream of tartar	2.5 mL
¼ tsp.	salt	1.2 mL
2¾ cups	sugar, divided	675 mL
2 cups	whipping cream	500 mL
1 - 8 oz. pkg.	cream cheese, softened	227 g
1 tsp.	vanilla	5 mL
2 cups	miniature marshmallows	500 mL
2 cups	rhubarb sauce (p. 2)	500 mL

Heat oven to 350°F. Grease a 13×9×2 inch (33×23×4 cm) baking pan.

In a large bowl, beat egg whites, cream of tartar and salt until foamy. Slowly beat in 1¾ cups (425 mL) sugar and continue beating until stiff and glossy. Do not under beat.

Spread meringue in greased baking pan and bake at 275°F for one hour. Turn off oven and leave meringue in with oven door closed for 12 hours or longer (overnight works well). In a chilled bowl, beat whipping cream until stiff. In another bowl, beat together softened cream cheese, vanilla and the remaining 1 cup (250 mL) sugar.

Gently fold whipped cream and marshmallows into cream cheese mixture. Spread over meringue.

Cover and chill for 12 to 24 hours.

Cut into serving sized pieces and top with rhubarb sauce. Makes 12 large or 16 medium servings.

Strawberry Flan with Rhubarb Sauce

Oven 375°F

A hint of Grand Marnier makes this dessert special.

Pastry:
2 cups	all purpose flour	500 mL
½ cup	sifted icing sugar	125 mL
1 tsp.	grated orange rind	5 mL
½ tsp.	baking powder	2.5 mL
1 cup	butter	250 mL

Filling:
12 ozs.	cream cheese	340 g
¾ cup	sifted icing sugar	175 mL
1½ tsp.	vanilla	7.5 mL
3 cups	fresh strawberries, halved	750 mL

Glaze:
1 cup	thickened cool rhubarb sauce (p. 2)	250 mL
2 tbsps.	sifted icing sugar	30 mL
3 tbsps.	Grand Marnier	45 mL
10 drops	red food coloring (optional)	10 drops

Heat oven to 375°F.

Pastry:
Cream butter. Gradually beat in flour, icing sugar, orange rind and baking powder.

 Well-known B.C. clay artist Bob Kingsmill uses rhubarb leaves in his work.

Heat oven to 375°F.

Pastry:
Cream butter. Gradually beat in flour, icing sugar, orange rind and baking powder.

Press evenly into an ungreased 11 inch (28 cm) recessed flan pan.

Prick well with a toothpick or a broom straw.

Bake for 15-20 minutes or until lightly browned. Invert over serving plate and gently unmold immediately after taking out of the oven. Cool completely.

Filling:
Beat cream cheese until soft and smooth. Beat in icing sugar and vanilla until mixture is creamy. Whip cream until stiff and fold into cream cheese mixture. Spread evenly in prepared crust. Chill until set, about 1 hour.

Glaze:
In a small bowl, stir together rhubarb sauce, icing sugar, Grand Marnier and food colouring. Set aside.

Wash and hull strawberries. Cut each in half lengthwise and arrange berries on top of cream cheese filling.

Spoon glaze over berries. Chill and serve. Serves 12.

To vary: orange juice may be substituted for Grand Marnier if desired.

Rhubarb Walnut Cheesecake with Cointreau Sauce

Oven 375°F

Crust:

1½ cups	graham cracker crumbs	375 mL
¼ cup	brown sugar	60 mL
¼ cup	finely chopped walnuts	60 mL
½ cup	melted butter or margarine	125 mL
1 cup	rhubarb walnut marmelade	250 mL

Filling:

1 lb.	cream cheese	454 g
¾ cup	granulated sugar	185 mL
2 tbsps.	flour	30 mL
2 tbsps.	Cointreau	30 mL
2	eggs	2

Cointreau Sauce:

½ cup	rhubarb walnut marmelade (p. 89)	125 mL
¼ cup	Cointreau	60 mL

Heat oven to 375°F.

Crust:
Stir together graham cracker crumbs, brown sugar and walnuts. Blend in melted butter. Press into the bottom of a greased 8 inch (22.5 cm) springform pan. Bake at 375°F for 8 minutes or until done. Cool

Heat oven to 375°F.

Crust:
Stir together graham cracker crumbs, brown sugar and walnuts. Blend in melted butter. Press into the bottom of a greased 8 inch (22.5 cm) springform pan. Bake at 375°F for 8 minutes or until done. Cool

Set oven to 350°F.

Spread 1 cup (250 mL) of rhubarb marmelade evenly over crust. Set aside.

Filling:
In a medium sized mixing bowl, cream together cream cheese, sugar, flour and Cointreau. Beat until well blended and smooth. Add eggs one at a time, beating well after each addition.

Pour mixture over marmelade covered crust. Bake at 350°F for 60 minutes or until cream cheese is firm.

Rest pan on wire rack and carefully run a knife around the rim to loosen cheesecake. Let cool completely before removing sides of pan, then chill cheesecake until ready to serve.

Sauce:
Gently heat remaining marmelade and Cointreau together in a small saucepan. Cool. Drizzle a tbsp. of sauce over each slice of cheesecake when set on individual serving plates. Serves 6-8.

 Actors will mutter Rhubarb, Rhubarb, Rhubarb to simulate indistinguishable crowd converstation.

Rhubarb Sponge Pudding Oven 325°F

¼ cup	water	60 mL
⅔ cup	granulated sugar	150 mL
6 cups	rhubarb cut into 1″ (2.5 cm) pieces	1500 mL
2 tbsps.	quick-cooking tapioca	30 mL
1 tbsp.	grated orange peel	15 mL
2	eggs	2
¼ tsp.	cream of tartar	1.25 mL
⅛ tsp.	salt	.62 mL
⅓ cup	icing sugar	75 mL
6 tbsps.	sifted cake flour	90 mL

Heat oven to 325°F. Butter a 6 cup (1.5 L) casserole or baking dish.

Combine water, rhubarb, granulated sugar, tapioca and orange peel in a large saucepan over moderate heat and bring to a full boil. Reduce heat and simmer for 15 minutes or until tender. Pour hot rhubarb mixture into baking dish.

Beat eggs in a small bowl until thick and creamy. Gradually add the icing sugar. Beat until sugar is dissolved between each addition. Sift flour over egg mixture. Fold through gently. Spread mixture evenly over hot rhubarb mixture.

Bake at 325°F for 35-40 minutes or until top is golden brown and set. When cool, sprinkle top with icing sugar. Serves 4-6.

Strawberry Rhubarb Puff Oven 450°F

1 - 16 oz.	package frozen rhubarb, thawed	454 g
1 - 10 oz.	package frozen strawberries, thawed	283 g
½ cup	granulated sugar	125 mL

Topping:

2 cups	all purpose flour	500 mL
2 tbsps.	sugar	30 mL
3 tsps.	baking powder	15 mL
1 tsp.	salt	5 mL
⅓ cup	salad oil	75 mL
⅔ cup	milk	150 mL
1½ tbsps.	butter or margarine	22 mL
2 tbsps.	sugar	30 mL
1 tsp.	cinnamon	5 mL

Heat oven to 450°F. In a greased 9×9×2 inch (23×23×5 cm) baking pan, mix rhubarb, strawberries, and ½ cup (125 mL) sugar. Place in oven until bubbly.

Measure flour, 2 tbsps. (30 mL) sugar, baking powder and salt into bowl. Pour oil and milk into measuring cup (do not stir together); pour all at once into flour mixture. Stir until mixture cleans side of bowl and forms a ball.

Drop dough by 9 spoonfuls onto hot fruit. Make an indentation in each biscuit and dot with butter. Mix 2 tbsps. sugar and cinnamon and sprinkle on the biscuits. Bake 20 to 25 minutes. Serve warm. 9 servings.

Fresh Strawberry Rhubarb Puff:
Substitute 3 cups fresh, cut up rhubarb and 1 pint fresh strawberries, sliced, for the frozen fruit. Increase the ½ cup (125 mL) sugar to 1½ to 2 cups (375-500 mL). Add ½ cup (125 mL) water to fruit mixture when cooking.

Rhubarb Floating Islands

2 cups	rhubarb sauce (p. 2)	500 mL
2	eggs, separated	1
⅛ tsp.	salt	.62 mL
¼ cup	sugar	60 mL
3 cups	milk	750 mL
3	eggs	3
½ cup	sugar	125 mL
¼ tsp.	salt	1.25 mL
1½ tsp.	vanilla	7.5 mL
½ tsp.	cinnamon	2.5 mL
¼ tsp.	almond extract	1.25 mL

Spread the 2 cups of rhubarb sauce in the bottom of serving dish. Beat 2 egg whites and ⅛ tsp. salt until soft peaks form. Gradually add ¼ cup sugar, beating until stiff peaks form. In a skillet, slowly heat 3 cups milk to simmer. Drop in meringue by tbsps. to make six. Cook slowly, uncovered, until firm, about 5 minutes. Lift from milk with a slotted spoon. Reserve milk. Drain meringues on paper towels.

To make custard layer, beat the 3 eggs and the egg yolks slightly. Add ½ cup sugar and salt. Stir in the 3 cups slightly cooled milk from meringues. Add more if needed to make up the 3 cups.

Cook over hot, not boiling water, stirring constantly until mixture coats a metal spoon. Remove from heat at once. Cool slightly and add vanilla.

Pour custard over rhubarb sauce in serving dish.

Top with meringues. Chill. Dust lightly with cinnamon for decoration. Serves 6.

Variation: Make with rhubarb orange sauce and grate a bit of orange rind over the finished desert for decoration.

Rhubarb Chiffon Parfait

A nice light dessert.

2½ cups	diced rhubarb	625 mL
½ cup	water	125 mL
¾ cup	sugar	175 mL
1 - 6 oz. pkg.	strawberry jello	170 g
3	egg whites	3
⅓ cup	sugar	75 mL

Combine rhubarb, water and sugar in medium saucepan and cook over medium heat until rhubarb is tender.

Remove from heat. Add strawberry jello and stir until dissolved. Chill until slightly thickened.

Beat egg whites until foamy. Add sugar and continue beating until mixture stands in soft peaks.

Place pan, with rhubarb mixture in it, in a bowl of ice and whip until fluffy and thick. Fold in egg whites. Turn into parfait glasses and chill until firm. Serve with whipped cream if desired. Serves 6.

To vary, try cherry flavoured jello.

Rhubarb Bavarian Cream

2	large eggs, separated	2
1 cup	milk	250 mL
1 envelope	unflavoured gelatin	1 envelope
⅛ tsp.	salt	.62 mL
¼ cup	sugar	60 mL
1 cup	heavy cream	250 mL
1 tsp.	vanilla	5 mL
1 cup	thickened, chilled rhubarb sauce (p. 2)	250 mL
	food colouring, if desired	

Lightly oil a 4 cup mold.

In a small saucepan, combine egg yolks, milk, gelatin and salt. Set aside 2 minutes to let gelatin soften.

Cook mixture over medium heat, stirring constantly, until mixture thickens and coats the spoon. Do not boil. Set aside for 5 minutes then refrigerate, checking occasionally, until the mixture mounds slightly when stirred.

Beat the egg whites until soft peaks form. Gradually add sugar, beating until stiff peaks form.

In a medium bowl, beat ¾ cup (185 mL) heavy cream and the vanilla until soft peaks form.

Fold egg whites and whipped cream into the gelatin mixture.

Spoon half of mixture back into a medium sized bowl. Fold remaining ¼ cup (60 mL) cream into one half of gelatin mixture.

Fold rhubarb sauce into other half of gelatin mixture.

Alternately spoon rhubarb gelatin mixture and vanilla gelatin mixture into the mold to form a marbled pattern.

Chill until firm, about 1½ hours.

To unmold, loosen edges of mold with a small spatula. Invert onto a serving plate. Cover mold with hot, damp cloth, Shake mold and remove. Refrigerate 5 to 10 minutes to set surface. Serves 6.

 A piece of rhubarb pie, double crust, contains about 400 calories.

Rhubarb Ginger Mousse

3 cups	rhubarb cut into 1″ (2.5 cm) pieces	750 mL
⅓ cup	granulated sugar	75 mL
½ cup	orange liqueur	125 mL
⅓ cup	water	75 mL
2 tbsps.	finely chopped crystallized ginger	30 mL
¼ tsp.	salt	1.25 mL
1 cup	confectioner's sugar	250 mL
1 envelope	unflavoured gelatin	1 envelope
2 tbsps.	cold water	30 mL
¼ cup	boiling water	60 mL
2 cups	whipping cream	500 mL

Combine rhubarb, granulated sugar, orange liqueur and salt in a heavy saucepan. Bring to boil stirring frequently. Reduce heat and simmer until mixture is the consistency of applesauce — (30 mins.).

Stir in ginger and simmer for a further 15 minutes. Add water if sauce is too thick. Remove from heat and cool. Stir in confectioner's sugar. Soak gelatin in cold water and dissolve. Add to boiling water. Stir into fruit mixture. Whip cream until soft peaks form. Gently fold into fruit and gelatin mixture. Gently spoon into serving goblets. Refrigerate covered for several hours. Serves 8.

Rhubarb Strawberry Mousse

3 cups	rhubarb cut into 1" (2.5 cm) pieces	750 mL
⅓ cup	granulated sugar	75 mL
½ cup	strawberry liqueur	125 mL
⅓ cup	water (if needed)	75 mL
¼ tsp.	salt	1.25 mL
1 cup	confectioner's sugar	250 mL
1 envelope	unflavoured gelatin	1 envelope
2 tbsps.	cold water	30 mL
¼ cup	boiling water	60 mL
2 cups	whipping cream	500 mL
10 drops	red food colouring	10 drops
½ cup	sliced, fresh strawberries	125 mL

Combine rhubarb, granulated sugar and strawberry liqueur in a heavy saucepan. Bring to a boil, stirring frequently. Reduce heat and simmer until mixture is the consistency of applesauce. Add water if sauce gets too thick. Remove from heat and cool. Stir in confectioner's sugar and salt.

Soak gelatin in cold water to dissolve. Add to boiling water. Stir into fruit mixture.

Whip cream until soft peaks form. Gently fold into fruit and gelatin mixture. Color with red food coloring. Spoon into serving dishes. Cover and refrigerate for several hours.

Before serving, garnish with fresh, sliced strawberries. Serves 8.

Rhubarb Cheesecake   Oven 350°F

Crust:

¾ cup	all purpose flour	175 mL
¼ tsp.	cinnamon	1.25 mL
¼ tsp.	baking soda	1.25 mL
¼ tsp.	salt	1.25 mL
¼ cup	packed brown sugar	60 mL
⅓ cup	butter	75 mL

Filling:

1 lb.	cream cheese, softened	454 g
⅔ cup	granulated sugar	150 mL
2 tbsps.	flour	30 mL
2 tbsps.	orange juice	30 mL
1 tbsp.	grated orange rind	15 mL
½ tsp.	vanilla	2.5 mL
2	eggs	2
1 cup	thickened rhubarb orange sauce (p. 4)	250 mL

Crust:

In a bowl, stir together flour, cinnamon, baking soda, salt and brown sugar. Cut in butter until mixture is crumbly.

Press into a greased 9 inch (25 cm) springform pan. Place on a baking sheet and bake for 10 minutes at 350°F until lightly browned and firm. Remove from oven and set on a wire rack.

Crust:

In a bowl, stir together flour, cinnamon, baking soda, salt and brown sugar. Cut in butter until mixture is crumbly.

Press into a greased 9 inch (25 cm) springform pan. Place on a baking sheet and bake for 10 minutes at 350°F until lightly browned and firm. Remove from oven and set on a wire rack.

Turn oven up to 450°F.

Filling:

Cream together cream cheese, sugar, flour, orange juice, orange rind and vanilla. Add eggs one at a time, beating well after each addition. Pour cream cheese mixture into pan on top of crust.

Spoon rhubarb sauce over cream cheese. Draw spoon lightly through both layers of filling, making sure you don't cut through the crust, for a marbled effect.

Bake in at 450°F oven for 10 minutes. Reduce heat to 250°F and bake for 20 minutes longer or until cream cheese filling is set.

Transfer pan to wire cooling rack. Carefully run a knife around the pan to loosen cheesecake. Let cool completely before removing sides of pan.

Chill before serving. Serves 8-10.

Rhubarb Peach Compote

1 - 15 oz. can	peach slices	425 g
2 cups	cut up rhubarb	500 mL
¼ cup	sugar	60 mL
¼ tsp.	nutmeg	1.25 mL

Drain peaches. Put rhubarb in a saucepan and add ¼ cup of the peach syrup. Add sugar and nutmeg. Cover and simmer about 5 minutes or until rhubarb is tender but still in pieces.

Put peach slices in serving dish. Pour rhubarb mixture over and stir lightly. Cool and then chill very well before serving. Serves 4.

 North Americans baked their first rhubarb pie somewhere in the early 1800's.

Honey Ginger Compote

3 cups	rhubarb	750 mL
¾ cup	honey	175 mL
2 tbsps.	cut up crystallized ginger	30 mL
	or	
1 tbsp.	cut up fresh ginger	15 mL

Double Orange Rhubarb Compote

1	medium peeled and chopped orange	1
3 cups	rhubarb cut into 1″ (2.5 cm) pieces	750 mL
¼ cup	Cointreau	60 mL
1 cup	sugar	250 mL

Method for both recipes above:

Microwave Directions:

Place all ingredients in a 2 quart casserole with lid slightly ajar and cook on high for 7 minutes, stirring once. Pour rhubarb into serving dish and chill well before serving.

Conventional Directions:
Put all ingredients in a medium sized saucepan. Cover and simmer until rhubarb is tender but still in pieces, 5-10 minutes. Add water if mixture gets too thick. Pour rhubarb into serving dish and chill well before serving. Serves 6.

Rhubarb Strawberry Compote

½ cup	water	125 mL
1 - 16 oz. pkg.	frozen rhubarb	454 g
	or	
2 cups	fresh rhubarb	500 mL
2 tbsps.	sugar	30 mL
2 cups	fresh strawberries	500 mL
¼ tsp.	ginger	1.25 mL

In medium saucepan, heat water to boiling. Add rhubarb, sugar and ginger. Cook for 5-10 minutes. Gently stir once or twice taking care to disturb shape of rhubarb pieces as little as possible. Wash and slice strawberries and place in a serving dish. Pour rhubarb over strawberries and gently stir to blend together.

Chill before serving. Serves 6.

Rhubarb Fool

2 cups	rhubarb cut into 1″ (2.5 cm) pieces	500 mL
¼ cup	sugar	60 mL
2 cups	milk	500 mL
1 tbsp.	sugar	15 mL
3	egg yolks	3
¼ tsp.	vanilla extract	1.25 mL
½ cup	whipping cream	125 mL
7-10 drops	red food colouring (optional)	7-10 drops

Put the fruit and sugar into a saucepan over very low heat and leave until soft; or cook in the top of a double boiler over boiling water.

Puree in blender or food processor, then cool.

Heat the milk and 1 tbsp. of sugar in double boiler until the sugar is dissolved. Take out ½ cup (125 mL) and cool slightly. Stir in the beaten egg yolks and gently pour back into top of double boiler. Add vanilla, cook over boiling water, stirring until thickened.

Beat the fruit puree and cooled custard together until well blended and smooth. Add the whipped cream and, if desired, the food colouring.

Pour into serving glasses and top with whipped cream. Serves 4.

Rhubarb Meringue Torte   Oven 225°F

Though very showy, this dessert is easy to prepare!

6	large egg whites	6
½ tsp.	cream of tartar	2.5 mL
1½ cups sugar		375 mL
3½ cups chilled, thickened rhubarb orange sauce (p. 4)		875 mL
	made with an orange liqueur	

Day before serving: In a medium sized bowl, with electric mixer, beat egg whites and cream of tartar until soft peaks form. Gradually add sugar, beating until sugar dissolves and stiff peaks form.

Heat oven to 225°F. Line 2 baking sheets with aluminum foil; grease foil. Using an 8 inch (20.5 cm) round baking pan for a guide, draw two 8-inch (20.5 cm) circles on one sheet, and one 8-inch circle on the other.

Spoon meringue into large pastry bag with star tip. Pipe a border inside the edge of each circle on baking sheets. Do not extend beyond circle. Meringue layers will expand during baking.

On one circle, pipe a second border inside the first. This will be your top circle.

Divide remaining meringue into center of other circles. Spread with a small spatula to fill centers and touch shell borders all the way around.

Bake meringues 3 to 4 hours or until crisp. Turn off oven. Leave meringues in oven several hours or overnight. If desired, cover centers of meringue with foil and place under the broiler to brown edges.

An hour before serving, lay one of filled circles on a serving plate.

Top with one third of rhubarb sauce. Repeat with second filled-in circle.

Top this with open meringue circle and spoon in remaining rhubarb sauce. Cover torte carefully and refrigerate until ready to serve. Serves 8-10.

 Rhubarb's laxative properties are still used today in homeopathic medicine.

Quick Rhubarb Cake in a Blanket

Oven 400°F

A quick, convenient, fancy dessert which looks like it takes a great deal more effort than it actually does!

1	7" (18 cm) bought sponge layer cake	1
¾ cup	thickened rhubarb sauce (p. 2)	175 mL
4	egg whites	4
¼ tsp.	cream of tartar	1.25 mL
1 cup	sugar	250 mL

Place cake on the center from a 9 inch (22.5 cm) springform pan or other heat-resistant round surface. Spread rhubarb sauce between layers of cake and on top.

Whip egg whites and cream of tartar together until peaks hold their shape. Gradually beat in sugar.

Spread meringue generously over and around cake.

Place in 400°F oven for 5 to 10 minutes or until meringue is golden brown and crisp.

Cool and serve. Serves 10.

To freeze rhubarb: wash stems after trimming away leaves and bottom part. Cut into 1 inch (2.5 cm) or ½ inch (1 cm) pieces. Put into plastic bags, seal and freeze. I find it very useful to freeze 4 cups (1L) to each bag as this makes about 2 cups (500 mL) rhubarb sauce. Another option is to cook up the sauce and then freeze this in premeasured amounts.

Frozen Delights

Rhubarb and Cointreau Sorbet

A light, crisp-tasting dessert.

3 cups	rhubarb cut into ½″ (1 cm) pieces	750 mL
1 cup + 2 tbsps.	sugar	280 mL
1½ cups	water	375 mL
⅓ cup	Cointreau	75 mL
1 tsp.	orange rind	5 mL
¼ cup	whipping cream	60 mL

Combine rhubarb, sugar, Cointreau, orange rind and ½ cup water in a medium saucepan. Simmer over low heat until rhubarb is tender and almost pureed, 15-20 minutes.

Process in blender or food processor until smooth. Stir in cream and remaining cup of water.

Freeze in an ice cream maker following manufacturer's instructions. Makes 1 quart (1 L).

Frozen Rhubarb Cream with Strawberry Coulis

A truly elegant dessert.

2 - 250 g pkgs.	cream cheese	500 g
½ cup	sour cream	125 mL
¾ cup	sugar	175 mL
2	egg yolks	2
2	egg whites	2
1½ cups	chilled rhubarb sauce (p. 2)	375 mL

Line a 9 × 5 inch (23 × 13 cm) loaf pan with double thickness of waxed paper.

Beat egg whites; set aside.

In a large bowl, beat cream cheese, sour cream, sugar and egg yolks until light and fluffy. Fold in beaten egg whites then rhubarb sauce. Pour into lined pan, smooth top. Cover and freeze at least 6 hours or until firm. If frozen hard, soften in refrigerator 1 hour before slicing.

To serve, remove from pan, unwrap and cut into 8 slices.

Spoon some strawberry coulis onto each dessert plate and top each with a slice. Serves 8.

Strawberry Coulis:

1 cup	fresh, sliced strawberries	250 mL
⅓ cup	sugar	75 mL
2 tbsps.	water	30 mL

In medium saucepan, combine berries, sugar and water. Bring to a boil. Cook and stir over medium heat for 2-3 minutes. Process in food processor or blender on high for 1 minute. Pass mixture through a fine sieve. Chill. Stir well before serving. Makes about 1 cup (250 mL).

Rhubarb Yogurt-sicles

1 cup	pureed rhubarb	250 mL
1 - 7 oz.	jar marshmallow creme	200 mL
2 cup	plain yogurt	500 mL
1½ cup	rhubarb cut into 1″ (2.5 cm) pieces	375 mL
¼ cup	water	60 mL

Wash rhubarb and remove any remaining leaf. Place rhubarb and water in a medium-sized saucepan. Over medium heat bring to a boil and simmer until rhubarb is tender, 5-10 minutes. Cool. Puree in a blender or food processor until smooth. Add marshmallow creme and blend until smooth. In a large bowl, stir yogurt until smooth. Gradually stir in rhubarb mixture.

Pour into ice cream maker or popsicle molds. Freeze according to manufacturer's instructions. Makes 1 quart or about 14 popsicles.

Freezer Method:
Pour prepared mixture into a 9 × 5 inch (23 × 13 cm) loaf pan. Cover with foil or plastic wrap. Freeze 3-6 hours or until firm. Stir with a fork 2 or 3 times while freezing. Makes 1 quart (1 L).

Rich Rhubarb Yogurt Creme

Strawberries and cream blended with rhubarb for a smooth taste delight.

1 - 3 oz. pkg.	strawberry flavoured gelatin	85 g
¾ cup	sugar	175 mL
1 cup	water	250 mL
2 cups	rhubarb sauce (p. 2)	500 mL
2 cups	plain yogurt	500 mL
1 cup	whipping cream	250 mL

In a medium saucepan, combine gelatin and sugar. Add water. Stir over medium heat until gelatin dissolves. Stir in rhubarb sauce and set aside.

In a medium bowl, stir yogurt until smooth. Stir in rhubarb mixture and whipping cream. Pour into ice cream cannister. Freeze in ice cream maker according to manufacturer's directions. Makes about 2 quarts (2 L)

 Rhubarb, like asparagus, is one of the very few perennial vegetables.

North South Ice Cream

The best of both north and south hemispheres combined in a cool summer treat.

2 cups	rhubarb sauce (p. 2)	500 mL
1 cup	crushed pineapple, drained	250 mL
1 tbsp.	coarsely grated orange peel	15 mL
2	eggs	2
1¼ cups	sugar	310 mL
2 cups	half and half cream	500 mL
1 cup	whipping cream	250 mL
¼ cup	crushed, drained pineapple (not pureed)	60 mL

Puree rhubarb sauce and drained, crushed pineapple in a blender or food processor. Stir in grated orange peel and set aside.

In a large bowl, beat eggs until thick and lemon coloured.

Whisk in sugar, half and half and whipping cream. Stir in fruit mixture and the remaining pineapple. Pour into ice cream maker. Freeze according to manufacturer's directions. Makes 2 quarts (2 L).

Frozen Rhubarb Custard

2 cups	rhubarb cut into 1" (2.5 cm) pieces	500 mL
¼ cup	water	60 mL
2 cups	half and half (light cream)	500 mL
3	eggs, beaten	3
1¼ cups	sugar	310 mL
1 cup	whipping cream	250 mL
1 tsp.	vanilla	5 mL

In a medium saucepan, combine rhubarb and water. Bring to boil. Reduce heat and simmer for 5-10 minutes, or until rhubarb is tender. Puree in blender or food processor. Cool.

In a medium saucepan, combine half and half, sugar and eggs. Cook over low heat, stirring constantly until mixture thickens and coats a metal spoon.

Stir in whipping cream, vanilla and pureed rhubarb. Pour into ice cream maker and freeze according to manufacturer's directions. Makes about 6 cups (1.5 L).

Rhubarb Sherbet

2 cups	rhubarb cut into 1" (2.5 cm) pieces	500 ml
⅓ cup + 1 tbsp.	honey	100 ml
	grated rind of ½ lemon	
1 cup	orange juice	250 ml
2	egg whites	2

Combine rhubarb, honey and lemon rind in a deep, heavy saucepan and cook over low heat until rhubarb is soft. Stir occasionally to prevent scorching. Puree in food processor or blender. Chill. Combine rhubarb puree and orange juice. Process in an ice cream maker until almost firm.

Beat egg whites to soft peaks and add to sherbet mixture.

Continue processing until firm.

Remove sherbet from drum and serve or store in freezer. Sherbet keeps well in freezer for 2 or 3 days. After that it will start to crystallize.

Yields: about 1 quart (1 L).

Rhubarb is a member of the buckwheat family.

Honey'n Rhubarb Ice Cream

3 cups	rhubarb	750 mL
1/4 cup	sugar	60 mL
1 tbsp.	grated lemon rind	15 mL
2	eggs	2
1/2 cup	honey	125 mL
1 tbsp.	water	15 mL
1 cup	whipping cream	250 mL
2 cups	half and half (light cream)	500 mL

Combine rhubarb, sugar, lemon rind and water in a deep, heavy saucepan. Cover and cook over low heat until rhubarb is soft, 5-10 minutes. Stir occasionally to prevent scorching. Puree in blender or food processor, then chill completely in freezer.

Beat eggs in mixer until light and frothy. Continue beating while slowly adding honey. Beat until mixture is thick.

Add heavy cream and light cream (half and half).

Pour into ice cream maker and process until almost firm. Add chilled rhubarb puree and continue processing until firm.

Remove paddle, pack down in drum or freezer container and harden in freezer.

Yields: about 2 quarts (2 L).

Ginger Snap and Honey Ice Cream

We found this really appealed to grown-up tastebuds.

3 cups	rhubarb	750 mL
¾ cups	honey	185 mL
2 tbsps.	cut up crystallized ginger	30 mL
2	eggs	2
1 cup	sugar	250 mL
2 cups	half and half (light cream)	500 mL
1 cup	whipping cream	250 mL
¼ cup	crushed ginger snaps	60 mL

Wash and cut up rhubarb. Put in medium sized saucepan with honey and ginger. Cook over medium heat until rhubarb is soft and saucey about 15 minutes. Chill. Puree in blender or food processor until smooth. In a large bowl, beat eggs together until thick and lemon coloured. Stir in sugar. Add in cream, half and half and rhubarb mixture.

Crush ginger snaps.

Process cream mixture in an ice cream maker according to manufacturer's directions. When ice cream is almost ready to serve, add cookie crumbs. Mix in well. Makes 2 quarts (2 L).

Strawberry Rhubarb Ice Cream

1 cup	stewed rhubarb or rhubarb sauce (p. 2)	250 mL
1 cup	sliced, fresh strawberries	250 mL
1 tbsp.	coarsely grated orange peel	15 mL
1 cup	sugar	250 mL
2	eggs	2
2 cups	half and half cream	500 mL
1 cup	whipping cream	250 mL

In a large bowl, beat eggs until thick and lemon coloured. Stir in sugar, half and half and whipping cream. Add rhubarb sauce, sliced strawberries and orange rind. Mix well.

Pour mixture into ice cream maker and freeze according to manufacturer's directions. Makes 2 quarts (2 L).

 "A rhubarb" was U.S. airforce slang in W.W.2 meaning a low-level strafing flight.

Rhuberry Ice

Refreshing as a spring breeze.

4 cups	fresh rhubarb	1000 mL
¼ cup	water	60 mL
1½ cups	sugar	310 mL
2 cups	fresh strawberries	500 mL

Combine rhubarb and water in a medium sized saucepan. Bring to a boil over medium heat. Cover and simmer for 5 minutes or until rhubarb is tender. Stir in sugar until dissolved. Puree rhubarb mixture in food processor or blender until smooth. Pour into a large bowl and set aside.

Wash and hull strawberries. Process in blender or food processor until almost smooth. Stir into rhubarb mixture. Place in an ice cream maker and process following manufacturer's directions for ice cream maker.

Freezer Method:
Pour fruit mixture into a 9 × 5 inch (23 × 13 cm) loaf pan. Cover with foil or plastic wrap. Freeze until firm, 3-6 hours.

Scrape frozen mixture with a fork until pieces resemble finely crushed ice.

Serve immediately.

Yields: about 5 cups (1.25 L).

Rhubarb Ice Cream

2 cups	rhubarb cut into 1" (2.5 cm) pieces	500 mL
1 cup	sugar	250 mL
	grated rind of ½ lemon	
3	eggs	3
2 cups	heavy cream	500 mL
2 cups	half and half or light cream	500 mL

In a deep, heavy saucepan, combine first three ingredients: rhubarb, sugar and lemon rind. Cover and cook over low heat until rhubarb is soft. Stir occasionally to prevent scorching. Puree in blender or food processor. Chill completely in freezer.

Beat eggs until light and frothy. Add heavy cream and light cream. Pour into ice cream maker and process until almost firm. Add chilled rhubarb puree and continue processing until firm.

Remove paddle, pack down in drum or freezer container and harden in freezer.

Yields: about 2 quarts.

Rhubarb Cheesecake Ice Cream

All the wonderful taste of cheesecake made into an ice cream, yum!

2 cups	rhubarb	500 mL
½ cup	sugar	125 mL
2 tbsps.	water	30 mL
1 tsp.	lemon rind	5 mL
2	eggs	2
1 cup	whipping cream	250 mL
2 cups	half and half	500 mL
8 ozs.	cream cheese	226 g
1 tsp.	vanilla	5 mL
¼ cup	graham cracker crumbs	60 mL
⅛ tsp.	cinnamon	.60 mL

Wash and chop rhubarb into ½ inch (1 cm) pieces. Simmer rhubarb, ⅓ cup (75 mL) sugar, and water together in a medium sized saucepan until rhubarb is saucey, soft, and falling apart. Stir occasionally to prevent scorching. Cool. Puree in a blender or food processor. Put in freezer and chill completely.

Blend together eggs and sugar. Add cream cheese and beat until smooth. Blend in whipping cream and vanilla. Stir in half and half.

Put cream cheese mixture into ice cream maker and process until quite firm.

Mix together graham cracker crumbs and cinnamon.

Spoon rhubarb sauce into ice cream maker. Mix into cream cheese mixture along with crumbs.

Finish processing. Makes 2 quarts (2 L).

Jams Jellies and Preserves

Rhubarb Ginger Jam

5 cups	diced rhubarb	1250 mL
¼ cup	chopped crystallized ginger	60 mL
1 tbsp.	powdered ginger	15 mL
6 cups	sugar	1500 mL
1 - 2 oz. pkg.	fruit pectin crystals	57 g

Combine rhubarb, chopped ginger, powdered ginger, sugar and pectin in a large, heavy saucepan.

Slowly bring to a boil, stirring frequently to prevent scorching. When the mixture has reached a full rolling boil (one that cannot be stirred down) maintain for one minute, stirring constantly. Still stirring, remove pot from heat, skim foam from top of jam with a metal spoon, discard.

Ladle jam into hot, sterilized jars. Fill to within ¼ inch, wipe around inside of rims of jars with a clean, damp cloth.

Seal jars with a layer of melted paraffin wax.

Yields: 5 or 6 medium-sized jars of jam.

Pineapple Rhubarb Jam

5 cups	diced rhubarb	1250 mL
5 cups	sugar	1250 mL
1½ cups	drained, crushed pineapple	375 mL
1 - 2 oz. pkg.	fruit pectin crystals	57 g

Place rhubarb and sugar in a large, heavy saucepan and slowly bring to a boil, stirring constantly. Cook until rhubarb is tender, about 25 minutes. Add pineapple and pectin. Bring to a boil, stirring constantly. Once mixture comes to a full rolling boil, hold for one minute stirring constantly. Remove from heat and skim off foam.

Seal in hot, sterilized jars.

Yields: about 5 or 6 medium jars.

*Gerarde's Herbal, written in 1597, specifies
"The best Rubarbe is that which is brought
from China fresh and newe...
The second in goodness is that which cometh
from Barbarie (North Africa)...
The last and woorst
from Bosphorus and Pontus (Turkey).*

Rhubarb Raspberry Jam

A tasty jam also an economical way to stretch your raspberry dollar.

5 cups	rhubarb	1250 mL
2 cups	fresh or frozen raspberries	500 mL
5 cups	sugar	1250 mL
1 - 2 oz. pkg.	fruit pectin crystals	57 g

Same directions as above.

Rhubarb Marmelade

A zesty, good morning taste.

6 cups	diced rhubarb	1500 mL
3	oranges	3
1 lemon		1
5 cups	granulated sugar	1250 mL
1 - 2 oz. pkg.	fruit pectin crystals	57 g

Remove seeds from oranges and lemons and chop fruit coarsely. Wash and cut up rhubarb. Thinly slice up lemon and orange, including peel. Cut these again in quarters. Mix fruit, sugar, water and pectin together in a large saucepan.

Slowly bring to a boil, stirring frequently. Once mixture is boiling, bring to a hard rolling boil and maintain, stirring constantly for one minute.

Remove pot from heat and skim off any foam with a metal spoon. Ladle marmelade into hot, sterilized jars.

Wipe around rims of jars with clean, damp cloth and seal with melted paraffin.

Yields: 5-6 jars of marmelade.

Rhubarb Walnut Marmelade

An unusual combination of flavours and textures that works extremely well.

8 cups	chopped rhubarb	2000 mL
5 cups	granulated sugar	1250 mL
2 cups	lemon, sliced thinly	500 mL
1 cup	orange, sliced thinly	250 mL
1 cup	chopped walnuts	250 mL
1 - 2 oz. pkg.	fruit pectin crystals	57 g

Chop and measure rhubarb and place in a large pot or preserving kettle. Add sugar, orange and lemon slices. Let stand for 24 hours.

Add pectin crystals.

Bring to boil and boil for 1 hour or until thickened, stirring frequently.

Add chopped walnuts. Pour into hot, sterilized jars and seal.

Yields: 4-6 medium jars.

Rhubarb Carrot Marmelade

A rich orange colour gives away this delicious marmelade's vegetable origin.

4 cups	rhubarb cut into 1" (2.5 cm) pieces	1000 mL
2 cups	grated carrots	500 mL
1	orange (medium sized)	1
1	lemon (medium sized)	1
5 cups	sugar	1250 mL
½ cup	water	125 mL
1 - 2 oz. pkg.	fruit pectin crystals	57 g

Wash and cut up rhubarb. Wash and grate carrots. Thinly slice up lemon and orange, including peel. Cut these again in quarters.

Mix fruit, sugar, water and pectin together in a large saucepan. Slowly bring to a boil, stirring frequently. Once mixture is boiling, bring to a hard rolling boil and maintain, stirring constantly for one minute.

Remove pot from heat and skim off any foam with a metal spoon. Ladle marmelade into hot, sterilized jars.

Wipe around rims of jars with clean, damp cloth and seal with melted paraffin.

Yields: 5-6 jars of marmelade.

Rhubarb Strawberry Conserve

5 cups	diced rhubarb	1250 mL
6 cups	sugar	1500 mL
2 cups	sliced strawberries	500 mL
1 - 2 oz. pkg.	fruit pectin crystals	57 g

Combine rhubarb and pectin in a heavy saucepan. Cook until mixture starts to boil. Add sliced strawberries. Continue to boil until a full rolling boil has been maintained for one minute, stirring constantly. Remove from heat and skim off foam with a metal spoon.

Ladle into hot, sterilized jars, filling to within ¼ inch of rim of jar.

Wipe edges with a clean, damp cloth.

Seal with melted paraffin wax.

Yields: 5 or 6 jars.

A cure for heartburn written in 1400 by Gov. Lordsh in his Secreta Secret lists "foure peny weght reubard" in his ingredients to "withdrawyse be fleume fro be mouth of be stomake."

Cinnamon Rhubarb Conserve

Sweet 'n sour taste with a secret ingredient.

6 cups	chopped rhubarb	1500 mL
⅓ cup	cinnamon heart candies	75 mL
½ cup	water	125 mL
¼ cup	white vinegar	60 mL
3 cups	granulated sugar	750 mL
1 tbsp.	lemon juice	15 mL
1 tbsp.	coarsely chopped ginger root	15 mL
3	whole cloves	3
2	whole allspice	2

In a large pot, combine rhubarb, cinnamon hear candies, water, vinegar, sugar and lemon juice.

Tie the ginger root, cloves and allspice together in a small muslin bag. Add to pot.

Slowly bring mixture to a boil, stirring frequently to prevent scorching. Simmer until desired thickness is reached.

Remove from heat and ladle into hot, sterilized jars and seal.

Serve with hot or cold meats.

Yields: 3 cups (750 mL) of conserve.

Spiced Rhubarb Jelly

An old fashioned accompaniment for duck and game or try it on hot cheese buns.

9 cups	rhubarb (2 cups juice, 500 mL)	2250 mL
5	whole cloves	5
1	2" (5 cm) cinnamon stick	1
5 cups	sugar	1250 mL
2	black peppercorns	2
6 ozs.	liquid fruit pectin	170 mL

Wash and chop rhubarb into 1 (2.5 cm) inch pieces. Process in a food chopper or blender until shredded.

Dampen a jelly bag or several thicknesses of cheesecloth. Spoon rhubarb into bag and hang from the knob of a cabinet over a medium-sized bowl. Let drip for several hours or until you are able to measure two cups of juice.

Tie spices into a small muslin bag and drop into juice. Let stand overnight. In the morning, pour juice and spices into a large pot. Add sugar and bring to a boil. Scoop out spices and add pectin. Bring to a full rolling boil (one that cannot be stirred down). Boil for 1 minute, stirring constantly.

Remove from heat. Continue stirring and skim off foam with a metal spoon. Pour into hot, sterilized jars.

Seal with melted paraffin wax.

Yields: 4½ cups (1125 mL) of jelly (5 jelly glasses).

Rhubarb Jelly

9 cups	rhubarb (2 cups juice, 500 mL)	2250 mL
5 cups	sugar	1250 mL
1 tbsp.	lemon juice	15 mL
6 ozs.	liquid fruit pectin	170 mL

Wash and chop rhubarb into 1 inch (2.5 cm) pieces. Process in a food blender or processor until pulped.

Hang pulp over a large bowl in a dampened jelly bag, or cheesecloth folded to a thickness of four. When you have two cups of rhubarb juice, discard pulp.

Pour juice, lemon juice and sugar into a large pot. Bring to a boil and add pectin. Bring to a full rolling boil for one minute, stirring constantly.

Remove from heat and skim off foam with a metal spoon.

Ladle jelly into hot, sterilized jelly glasses.

Wipe the edges of the jelly glasses with a clean, damp cloth.

Seal with a layer of melted paraffin wax.

Yields: 5 jelly glasses.

Rhuberry Jelly

8 cups	rhubarb cut into ½" (1 cm) pieces	2000 mL
½ cup	undiluted strawberry juice concentrate	125 mL
5 cups	sugar	1250 mL
6 ozs.	liquid fruit pectin	170 mL

Wash and chop rhubarb. Process through blender or food processor. Hang pulp in a jelly bag or 4 layers of cheesecloth and allow to drain until you have 1½ cups of rhubarb juice.

Add strawberry juice concentrate, rhubarb juice and sugar together in a large pot. Slowly bring to boil, stirring frequently.

When the syrup is boiling, add pectin. Stir constantly. Allow mixture to come to a full rolling boil for 1 minute.

Remove from heat still stirring. Skim foam off top of jelly and pour into hot, sterilized jelly glasses.

Yields: 6 jelly glasses.

Rhubarb and Apricot Chutney

8 cups	rhubarb, tops removed	2000 mL
2 cups	firmly packed dark brown sugar	500 mL
1 cup	cider vinegar	250 mL
1 cup	chopped dried apricots	250 mL
1 cup	finely chopped onion	250 mL
1 cup	dark seedless raisins	250 mL
2 tsp.	grated lemon rind	10 mL
2 tsp.	ground ginger	10 mL
½ tsp.	dry mustard	2.5 mL
¼ tsp.	ground chilis	1.25 mL
¼ tsp.	allspice	1.25 mL
½ tsp.	salt	2.5 mL

Trim off ends and any remaining leaves from rhubarb; wash stalks carefully, coarsely chop.

In a 3-quart saucepan, combine brown sugar, vinegar, apricots, onion, raisins, lemon rind, ginger, mustard, chilis, allspice and salt. Heat to boiling. Cook uncovered over high heat 5 minutes, stirring constantly. Stir in rhubarb, reduce heat and let simmer 1 hour or until rhubarb is just tender and saucey.

Ladle thickened chutney into hot sterilized sealers.

Store for 6 to 8 weeks before serving as flavour mellows and blends with time.

Yields: 4-6 pints (2000-3000 mL).

Rhubarb Raisin Chutney

2 qts.	rhubarb cut into 1″ (2.5 cm) pieces	2 L
2 cups	minced onion	500 mL
3 cups	cider vinegar	750 mL
4 cups	light brown sugar	1000 mL
1 tbsp.	fresh ginger	15 mL
1 tsp.	pepper	5 mL
1 tsp.	salt	5 mL
1 tbsp.	allspice	15 mL
½ tbsp.	cinnamon	7.5 mL
½ tbsp.	cloves	7.5 mL
½ tbsp.	mustard seeds	7.5 mL
2 cups	apples	500 mL

Wash and trim rhubarb. Cut up. Wash, peel and slice apples. Peel and mince onion. Combine rhubarb, apples, raisins, vinegar and onion in a large, heavy saucepan. Cook 20 minutes. Add sugar and spices (tied in a bag) and cook until thick, about 45 minutes, stirring constantly.

Seal in hot, sterilized jars.

Yields: 4-6 pints (2000-3000 mL).

Rhubarb Chutney
Made with Dates and Tomatoes

8 cups	rhubarb cut into 1" (2.5 cm) pieces	2000 mL
5 cups	stoned dates, cut up coarsely	1250 mL
2 cups	onions	500 mL
1¼ cups	chopped tomatoes	310 mL
1 tsp.	salt	5 mL
2 tbsps.	mustard seed	30 mL
1 tbsp.	ground ginger	15 mL
1 tsp.	ground chili flakes	5 mL
¼ cup	brown sugar or brown sugar to taste	60 mL
2 cups	cider vinegar	500 mL

Combine all ingredients in large, heavy saucepan and simmer to desired consistency. Stir frequently to prevent scorching. Add water during cooking when chutney gets too thick, about 3 hours.

Pour into hot, sterilized jars and seal.

Yields: 5 pints (10 cups) (2500 mL).

Rhubarb Fig Preserves

Thick and rich! Almost like being allowed to spoon raisin pie filling out of the jar.

4 cups	rhubarb	1000 mL
4 cups	granulated sugar	1000 mL
1 cup	chopped figs	250 mL
¼ cup	lemon juice	60 mL
½ cup	raisins	125 mL
1 tsp.	salt	5 mL

Combine the ingredients in a large pot, cover and let stand 24 hours.

Bring to boil and cook rapidly until the jellying point is reached. Stir frequently to prevent scorching.

Pour into hot, sterilized jars and seal.

Yields: about 4 cups (1000 mL).

Rhubarb Onion Relish

The rich tangy aroma of this relish filled our home every summer.

4 cups	diced rhubarb	1000 mL
2 cups	cider vinegar	500 mL
1 tbsp.	salt	15 mL
1 tbsp.	whole pickling spice	15 mL
4 cups	sliced onions	1000 mL
3½ cups	brown sugar	875 mL
½ tsp.	pepper	2.5 mL

Place pickling spice in a bag and cook with rest of ingredients in heavy saucepan, simmering for about 2 hours, or until onions are transparent and rhubarb saucey. Stir frequently. Seal in hot, sterilized jars. Yields: 2 pints.

In Act 5, Scene 3, Shakespeare has Macbeth railing at the Doctor who dared suggest he might heal himself, "What Rhubarb, Cyme, or what Purgative drug would scour these English hence?"

Pies and Tarts

Rhubarb Pie   Oven 425°F

A new gingery twist to an old fashioned favourite!

3½-4 cups	chopped rhubarb	875-1000 mL
5 cups	boiling water	1250 mL
2	eggs, separated	2
1 tbsp.	melted butter	15 mL
1 cup	granulated sugar	250 mL
¼ tsp.	salt	1.25 mL
1 tbsp.	cornstarch	15 mL
3 tbsps.	cold water	45 mL
1 tbsp.	finely chopped candied ginger	15 mL
1	unbaked 9″ (23 cm) pie shell	1
¼ cup	granulated sugar	60 mL

Cover rhubarb with boiling water and let stand 5 minutes. Drain well. Beat the egg yolks; add melted butter. Combine the 1 cup sugar, salt, cornstarch and water; blend into egg yolk mixture. Add drained rhubarb and chopped ginger.

Pour into pie shell. Make lattice top with strips of leftover pastry, leaving about 1½ inches between strips.

Bake at 425°F for 10 minutes. Reduce heat to 350°F and bake a further 35 minutes.

Make a meringue with egg whites and the ¼ cup sugar. Spoon meringue into the spaces between the lattice strips. Return to the oven until delicately brown, approximately 8 minutes. Serves 6-8.

Rhubarb Chiffon Pie

Oven 375°F

Crust:

1¼ cups	graham wafer crumbs	310 mL
¼ cup	finely chopped pecans	60 mL
½ cup	butter, melted	125 mL
¼ cup	brown sugar	60 mL

Melt butter. In a small bowl, combine graham wafer crumbs, pecans and brown sugar. Pour in butter and mix until crumbly. Pat into a 10 inch (25 cm) pie plate and bake at 375°F for 5-8 minutes. Cool completely before filling.

1¼ cups	cool rhubarb sauce (p. 2)	310 mL
¼ cup	sugar	60 mL
2 tbsps.	gelatin	30 mL
¼ cup	cold water	60 mL
½ cup	hot water	125 mL
¼ tsp.	salt	1.25 mL
½ cup	whipping cream	125 mL
2	egg whites	2
¼ cup	sugar	60 mL

Mix the salt and sugar into the rhubarb sauce.

In a large bowl, soften gelatin in cold water for 5 minutes. Pour in hot water and stir to dissolve the gelatin completely. Cool.

Whip cream until peaks form. Add rhubarb sauce to gelatin mixture. Gently fold in whipped cream.

Beat egg whites until soft peaks form. Add sugar, continue beating until stiff peaks form. Fold into rhubarb mixture.

Pour into crust. Chill until firm. Top with whipped cream if desired. Serves 6

Rhubarb and Raisin Pie   Oven 400°F

A lovely combination of textures and flavours.

3½ cups	rhubarb cut into ½" (1 cm) pieces	875 mL
½ cup	raisins	125 mL
1½ cups	sugar	375 mL
¼ tsp.	salt	1.25 mL
1 tbsp.	lemon juice	15 mL
1	egg, beaten	1
1	unbaked double crust pastry shell	1
¼ cup	flour	60 mL

Heat oven to 400°F. Line pie plate with pastry. Combine rhubarb, raisins, sugar, lemon juice, flour and beaten egg in a bowl and mix well.

Place in a pie shell and top with remaining pie dough.

Bake at 400°F for 10 minutes. Reduce heat to 350°F and bake until crust is golden brown and fruit is bubbly. Serves 6-10.

Sour Cream Rhubarb Pie   Oven 425°F

The creme de la creme of rhubarb pies.

1½ cups	chopped rhubarb	375 mL
1	unbaked 9″ (23 cm) pie shell	1
1 cup	brown sugar	250 mL
¼ cup	flour	60 mL
1 cup	sour cream	250 mL
3	eggs, separated	3
6 tbsps.	granulated sugar	90 mL

Place rhubarb in an unbaked shell. Mix brown sugar and flour well. Blend in sour cream until smooth. Add well-beaten egg yolks; pour over rhubarb.

Bake at 425°F for 10 minutes. Then turn oven down to 350°F and bake for 40 minutes. Remove pie from oven. Cool slightly. Reset oven to 425°F.

Beat egg whites until foamy. Gradually add sugar until egg whites form stiff, straight peaks.

Cover pie with meringue. Return to the oven until meringue is golden brown, 3-5 minutes. Serves 6-8.

Rhubarb Crumb Top Pie   Oven 400°F

Topping:

⅓ cup	all purpose flour	75 mL
⅓ cup	granulated sugar	75 mL
½ tsp.	cinnamon	2.5 mL
¼ tsp.	allspice	1.25 mL
3 tbsps.	butter or margarine	45 mL
1	unbaked 9" (23 cm) pastry shell	1

Filling:

3 cups	rhubarb cut into ½" (1 cm) pieces	750 mL
¼ cup	flour	60 mL
1	large egg	1
1 tsp.	vanilla	5 mL
1 cup	sugar	250 mL

Heat oven to 400°F. Trim off ends and any remaining leaves from rhubarb. Wash stalks carefully. Cut into ½ inch pieces.

In a medium sized bowl, beat egg until frothy. Beat in granulated sugar, ¼ cup (60 mL) flour, and the vanilla. Fold in the rhubarb. Spoon rhubarb mixture into pie crust.

In a small bowl, combine flour, sugar, cinnamon and allspice. Cut in butter with a fork until mixture resembles coarse crumbs. Sprinkle evenly over rhubarb filling.

Bake pie at 400°F for 10 minutes and then reduce heat to 350°F. Bake for another 35 minutes or until golden brown and bubbly. Cool to room temperature on wire rack before serving. Serves 6.

Almost Rhubarb Pie

Oven 350°F

A delicious cross between a square and a pie.

1½ cups	flour	375 mL
3 tbsps.	confectioner's sugar	45 mL
½ cup	butter or margarine	125 mL
1 cup	sugar	250 mL
¾ tsp.	salt	3.75 mL
2	eggs, beaten	2
2 cups	finely chopped rhubarb	500 mL

Mix 1 cup flour, sugar and butter until crumbly. Press into pie plate.

Bake at 350°F for 15 mins.

Mix together sugar, salt, beaten eggs and ½ cup flour. Stir in rhubarb. Spoon into prepared crust.

Bake at 350°F for 35 mins. or until set. Serves 8.

 Rhubarb leaves are poisonous. They contain a high level of oxalic acid.

Rhubarb and Raisin Custard Pie

Oven 450°F

Filling:

¼ cup	raisins	60 mL
1 tbsp.	grated orange rind	15 mL
1¼ cup	granulated sugar	310 mL
¼ cup	all purpose flour	60 mL
5 cups	chopped rhubarb	750 g
2	eggs, beaten lightly	2
2 tbsps.	melted butter or margarine	30 mL
1	double 9″ (23 cm) unbaked pie crust	1 mL

Heat oven to 450°F.

Blend together sugar and flour. Sprinkle one quarter of mixture over bottom of pie shell. Mix together rhubarb, raisins and orange rind. Put in pie shell. Mix together remaining flour and sugar with eggs and butter. Spread evenly over rhubarb. Top with a lattice crust.

Bake in a 450°F oven for 10 minutes and then reduce oven to 350°F and bake for 55 to 60 minutes longer or until filling is set and crust is golden brown. Makes 6-8 servings.

Apple Rhubarb Pie Oven 425°F

Perk up Mom's apple pie with a tangy rhubarb flavour.

Pastry for a 9″ (23 cm) two crust pie

1¼ cups	sugar	310 mL
¼ cup	flour	60 mL
1 tsp.	cinnamon	5 mL
¼ tsp.	cloves	1.25 mL
3 or 4	apples, pared, cored and sliced, about 3 cups	750 mL
2½ cups	rhubarb cut into 1″ (2.5 cm) pieces	625 mL
2 tsps.	margarine or butter	10 mL

Divide dough in half. Roll out half the dough and line a 9 inch (23 cm) pie plate. Trim edges.

In a large bowl, combine sugar, flour, cinnamon and cloves; mix well. Stir in rhubarb and apple. Turn into pastry lined pie plate. Dot top with margarine.

Roll out remaining pie dough. Moisten edge of bottom crust with water. Place pastry over pie. Seal and flute edges. Cut vents in top.

Bake at 425°F for 20 mins. Reduce heat to 350°F and continue baking for another 20 to 30 minutes or until top is golden brown and fruit is soft.

Rhubarb Cream Pie

Oven 350°F

1 tbsp.	melted butter or margarine	15 mL
2	eggs, separated	2
2 cups	diced rhubarb	500 mL
1 cup	sugar	250 mL
⅛ tsp.	salt	.62 mL
2 tbsps.	cornstarch	30 mL
¼ cup	whipping cream or canned cream	60 mL
¼ tsp.	nutmeg	1.25 mL

Mix melted butter, rhubarb, nutmeg and ½ cup sugar together in saucepan and cook slowly until the rhubarb is tender.

Combine remaining ¼ cup sugar with cornstarch, cream, salt and well-beaten egg yolks.

Add this to the rhubarb mixture and continue cooking until it thickens to a pudding consistency.

Cool, then pour into the baked pie shell.

Beat the egg whites until they form peaks and fold in ¼ cup sugar. Cover the pie with meringue and bake in a 350°F oven until the meringue is golden brown.

To keep an empty pie shell flat in pie plate while baking, prick with a fork and top with a layer of dried white beans, then bake. When done, pour out beans and continue on with your recipe.

Rhubarb-Strawberry Deep Dish Pie

Oven 400°F

1 - 16 oz. pkg.	fresh strawberries	454 g
	or	
2 cups	sliced, fresh strawberries	500 mL
4 cups	rhubarb cut into ½" (1 cm) pieces	1000 mL
2 tsps.	grated lemon rind	10 mL
1 cup	sugar	250 mL
½ cup	flour	125 mL
1 tbsp.	butter	15 mL
	pastry for one crust pie	
1½ tsps.	milk	7.5 mL
1 tbsp.	sugar	15 mL

Heat oven to 400°F. Butter an 8×8 inch (20.5×20.5 cm) glass baking dish.

Wash and slice strawberries or thaw strawberries enough to break apart. Wash and chop rhubarb.

Combine rhubarb, strawberries and lemon rind lightly in a large bowl. Sprinkle with lemon juice. Combine sugar and flour and toss lightly through the fruit. Put in the prepared baking dish. Dot with butter.

Roll pastry in a square larger than the baking dish. Lay it over the fruit, turn the edges of the pastry under and crimp and seal them well to the sides of the dish. Cut vents in the top to let steam escape. Brush the pastry with milk and then sprinkle it generously with sugar.

Bake about 45 mins. or until the pastry is well browned and filling is bubbling well. Serves 6.

Rhubarb Custard Pie   Oven 425°F

2½ cups	rhubarb cut into ½" (1 cm) pieces	625 mL
¼ cup	flour	60 mL
1 cup	granulated sugar	250 mL
2	eggs, separated	2
1 tbsp.	melted butter or margarine	15 mL
1	unbaked 9" (23 cm) pastry shell	1

Mix the sugar with slightly beaten egg yolks. Stir in the flour and rhubarb and blend well.

Add melted butter or margarine. Pour into pastry shell and bake at 425°F for about 10 minutes. Lower heat to 325°F for about 30 minutes.

Make meringue by beating egg whites stiff with ½ cup sugar. Spread on top of pie.

Return to oven and bake for 15 to 20 minutes or until golden brown. Serves 6-8.

Rhubarb Crumb Tart Oven 350°F

1½ cups	flour, divided	375 mL
1 tsp.	baking powder	5 mL
1 tsp.	salt	5 mL
½ cup	butter	125 mL
1	egg, beaten	1
1 tbsp.	milk	15 mL
3 cups	raw rhubarb, diced	750 mL
1 - 3 oz. pkg.	dry strawberry flavoured gelatin	1 - 85 g
1 cup	sugar	250 mL

Mix 1 cup flour, baking powder, salt and butter as for pie crust. Add egg and milk. Pat into an 8×10×2 inch pan. Place rhubarb in crust. Sprinkle gelatin over rhubarb. Mix 1 cup sugar, remaining flour and remaining butter together. Sprinkle over rhubarb mixture.

Bake 45 mins. at 350°F.

Serve with whipped cream. Serves 12.

Rhubarb Cinnamon Tart   Oven 400°F

Bright, cheery red hearts add a playful touch to this sophisticated tart.

2 cups	rhubarb	500 mL
¼ cup	sugar	60 mL
2 tbsps.	flour	30 mL
½ tsp.	cinnamon	2.5 mL
½ cup	heavy cream	125 mL
¾ recipe	standard pie pastry	¾
1 tbsp.	cinnamon hearts	15 mL

On a lightly floured surface, roll out the pastry to a thickness of ⅛ inch (.25 cm) and use it to line an 8 inch (20.5 cm) flan ring set on a greased baking sheet.

Trim and wash rhubarb. Cut into ½ inch (1 cm) pieces diagonally. Arrange neatly on the pastry.

Mix the sugar, flour, cinnamon and cream together and spoon over the rhubarb.

Put the tart on the middle shelf of an oven pre-heated to 400°F and bake for 30-40 mins. Cool for 10 mins. then sprinkle with cinnamon hearts.

If the tart is to be served warm, let it rest for 10 mins., otherwise it will be difficult to cut. Serves 8.

Rhubarb Butter Tarts　　　　　　　Oven 425°F

Everyone from the very old to the very young has enjoyed these tarts.

1 recipe	double crust pie pastry	1
½ cup	light brown sugar	125 mL
½ cup	corn syrup	125 mL
¼ cup	butter or margarine, melted	60 mL
1	egg, slightly beaten	1
1 tsp.	vanilla	5 mL
¼ tsp.	salt	1.25 mL
2 tbsps.	oatmeal	30 mL
1 cup	rhubarb cut into ½" (1 cm) pieces	250 mL

Heat oven to 425°F. Roll out pastry and line 12 tart shells. Melt and slightly cool butter. Combine brown sugar, syrup, butter, egg, vanilla, salt and oatmeal.

Evenly divide rhubarb pieces between tart shells.

Spoon liquid mixture over rhubarb. Fill tarts ⅔ full.

Bake on bottom shelf of oven at 425°F for 12 to 15 minutes or until just set.

Cool on wire rack, then remove from pans. Makes 12 tarts.

Rhubarb Strawberry Tarts   Oven 425°F

1 recipe	double crust pie pastry	1
1 cup	rhubarb cut into ½" (1 cm) pieces	250 mL
¾ cup	strawberry jam	185 mL
¼ cup	flour	60 mL
2 tbsps.	granulated sugar	30 mL

Heat oven to 425°F. Roll out pastry and line 12 tart shells.

Stir together jam and flour. Mix in rhubarb. Spoon mixture into tart shells.

Bake at 425°F for 20 to 25 minutes or until filling is bubbly and rhubarb is tender.

Cool on wire rack for 4 or 5 minutes and then sprinkle with sugar. Makes 12 tarts.

Rhubarb Custard Tarts   Oven 350°F

1 recipe	double crust pie pastry	1
1½-2 cups	rhubarb cut into ½" (1 cm) pieces	500 mL
1 cup	light brown sugar	250 mL
2	eggs	2
¼ tsp.	salt	1.25 mL
2 tbsps.	melted margarine or butter	30 mL

Heat oven to 350°F. Roll out pastry and line 12 tart shells.

Melt butter. Divide rhubarb pieces between tart shells.

Mix together eggs, brown sugar, salt and margarine. Spoon over rhubarb.

Bake at 350°F for 12-15 minutes or until filling is set and rhubarb is tender. Makes 12 tarts.

Bits and Pieces

Scandinavian Style Rhubarb

8 cups	rhubarb cut into ½″ (1 cm) pieces	2000 mL
½ cup	water	125 mL
1 cup	sugar	250 mL
3 tbsps.	cornstarch	45 mL
½ tsp.	salt	2.5 mL

Combine rhubarb and water in saucepan. Cover and cook until the rhubarb is breaking up into sauce.

Pour into a strainer set over a large measuring cup.

Press pulp until you have 2 cups of juice. Discard pulp. Combine sugar, cornstarch and salt in a saucepan. Stir in rhubarb juice gradually, blending well. Set over moderate heat and cook until boiling, stirring constantly. Boil for one minute.

Spoon into serving dishes. Chill well and serve with heavy cream or yogurt.

To vary, use ¼ cup (60 mL) fruit juice added to 1¾ cups (435 mL) rhubarb juice. Strawberry or orange juice is especially nice. Serves 8.

Another name for rhubarb is pie plant.

Rhubarb and Strawberry Tapioca

Vanilla pudding adds a creamy touch to blended taste of strawberry & rhubarb.

2 cups	rhubarb cut into ½" (1 cm) pieces	500 mL
2 cups	whole frozen strawberries, or fresh	500 mL
1 cup	water	250 mL
1 tbsp.	lemon juice	15 mL
¾ cup	sugar	185 mL
½ tsp.	salt	2.5 mL
⅓ cup	tapioca	80 mL
1 - 6 oz. pkg.	instant vanilla pudding (optional)	170 g

Wash and trim rhubarb. Cut into ½ inch pieces. Wash and hull strawberries. Put water, strawberries, rhubarb and lemon juice in a large saucepan.

Simmer fruit for five minutes.

Add sugar, salt and tapioca. Continue simmering until tapioca is translucent and soft and fruit mixture has thickened.

Spoon into individual serving dishes and chill until set about 2-3 hours. Top with ice cream or whipped cream to serve or pour tapioca into a large serving dish. Cool.

Mix a package of instant vanilla pudding and pour on top of tapioca. Chill. Top this with whipped cream to serve.

Rhubarb Barbecued Short Ribs

Oven 350°F

4 lbs.	beef short ribs, cut into pieces	2 kg
1 tsp	chili powder	5 mL
1	leak	1
½ tsp.	salt	2.5 mL
1 clove	garlic	1
¼ tsp.	ground black pepper	1.25 mL
½ cup	water	125 mL

Barbecue Sauce:

1 cup	rhubarb cut into 1" (2.5 cm) pieces	250 mL
½ cup	brown sugar	125 mL
2 tbsps.	cider vinegar	30 mL
2 tbsps.	catsup	30 mL
2 to 4 drops	hot red-pepper sauce	2 to 4 drops
½ tbsp.	prepared mustard	7.5 mL

Heat oven to 350°F. Trim excess fat from short ribs. In a 6 quart Dutch oven, brown short ribs on all sides. Sprinkle with chili powder, salt and pepper; add leek, garlic clove and water.

Cover and bake 1½ hours or until tender. Remove ribs to a roasting pan with a rack. Discard liquid in pot.

Barbecue Sauce:
In a medium sized saucepan, combine rhubarb, brown sugar, cider vinegar, catsup, hot red-pepper sauce and mustard. Slowly heat to boiling. Boil sauce for 5 minutes, stirring frequently.

Brush some of rhubarb mixture over ribs. Bake 15 minutes. Brush more mixture over ribs; bake until glazed and lightly browned — 15 to 20 minutes longer.

Serve with some extra barbecue sauce.

Try this sauce on poultry or white fish as well.

Chilled Rhubarb Soup

A delightfully fruity cold soup.

6 cups	rhubarb cut into 1" (2.5 cm) pieces	1500 mL
⅓ cup	sugar	75 mL
½ tsp.	salt	2.5 mL
1 - 10 oz. pkg.	frozen raspberries, thawed	283 g
½ cup	ruby port	125 mL
1 tbsp.	cornstarch	15 mL
¼ cup	plain or vanilla yogurt	60 mL

In a large saucepan, combine rhubarb, sugar and salt. Heat to boiling over high heat, stirring frequently. Reduce heat and cover. Simmer until mixture is saucey.

Stir raspberries into rhubarb sauce. Heat to boiling.

In a small bowl, combine port and cornstarch. Stir into fruit mixture. Cook until soup is clear and thickened, stirring constantly. Puree the thickened soup in a blender or food processor. Strain into a glass bowl. Cover and refrigerate until cold.

To serve, divide soup into individual portions. Stir yogurt to liquify and drizzle gently over soup to garnish. Serves 4-6.

Rhubarb Fruit Leather

9 cups	rhubarb	2250 mL
1½ cups	sugar	375 mL
1 cup	water	250 mL

Wash and chop rhubarb. Put in a large pot along with sugar and water Simmer until mixture is smooth. No chunks of rhubarb remaining Cool.

Set oven very low. Line 3 cookie sheets with wax paper or plastic wrap

Pour ⅓ of the rhubarb sauce onto each sheet and spread evenly. Pu in oven, taking care that cookie sheet liner doesn't come into contac with the stove element. Leave until rhubarb easily lifts away from pape (is leathery) or dry in dehydrator according to manufacturer's direc tions. Roll up in liner and store in freezer or a cool, dry place.

To vary, cut amount of rhubarb and make up the difference with apples pears, strawberries, etc.

 Rhubarb's most popular use from the 1400's to the 1700's was medicinal. The root was used as a very strong purge.

Rhubarb and Beets Harvard Style

A new approach to an old favourite.

1 cup	rhubarb	250 mL
1 tbsp.	water	15 mL
1 tbsp.	cornstarch	15 mL
3 tbsps.	sugar	45 mL
1 tbsp.	white vinegar	15 mL
⅛ tsp.	salt	.62 mL
2 tbsps.	marmelade	30 mL
2 cups	cooked, sliced beets	500 mL

Trim off ends and any remaining leaves from rhubarb. Wash stalks and cut into ½ inch (1 cm) pieces.

In a medium sized saucepan, combine rhubarb, sugar, vinegar and salt. Cook rhubarb until saucey... no large, separate pieces of rhubarb remain. Put beets on to heat slowly in a separate pot.

Mix cornstarch and cold water together and stir into rhubarb sauce. Continue cooking until cornstarch clears and sauce thickens.

Remove from heat and stir in marmelade.

Drain beets and gently stir in rhubarb sauce. Pour into a serving bowl and serve hot. Serves 6.

Rhubarb Cream Cheese Jellied Salad

2 cups	rhubarb diced	500 mL
½ cup	sugar	125 mL
¼ cup	water	60 mL
1 - 6 oz. pkg.	strawberry flavoured gelatin	170 g
1 tbsp.	lemon juice	15 mL
1¾ cups	cold water	435 mL
4 ozs.	cream cheese, softened	113 g
¼ cup	finely chopped celery	60 mL
1	banana, sliced	1

In a saucepan, combine rhubarb, ½ cup sugar and water. Cook covered over medium heat until rhubarb is just tender, about 3 minutes. Stir occasionally. Cool. Drain, reserving syrup.

Add enough water to syrup to make 2 cups. Return to saucepan. Bring to boil and stir in strawberry gelatin until dissolved. Add cold water and the lemon juice. Chill until partially set.

Add rhubarb, cream cheese, celery and banana. Turn into a 6 cup (1500 mL) mold. To serve, unmold onto serving plate. Serve with Fluffy Lemon Orange Dressing (following recipe). Serves 6-8.

Fluffy Lemon-Orange Dressing

1	egg	1
1 tsp.	grated lemon peel	5 mL
1 tbsp.	grated orange peel	15 mL
2 tbsps.	lemon juice	30 mL
⅓ cup	sugar	75 mL
1 cup	whipping cream	250 mL
½ tsp.	salt	2.5 mL

In saucepan, beat egg, add lemon and orange peel, the 2 tbsps. lemon juice, salt and ⅓ cup sugar.

Cook and stir over low heat until thickened, about 5 minutes. Cool to room temperature.

Whip cream. Fold into citrus mixture.

Chill.

Makes about 2 cups (500 mL).

Jellied Rhubarb Strawberry Salad

A pretty, low-calorie luncheon dish. Lovely served with cottage cheese.

1 - 14 oz. can	pineapple tidbits	498 mL
1 - 6 oz. pkg.	pineapple flavoured gelatin	170 g
1 cup	rhubarb cut into ½" (1 cm) pieces	250 mL
3 cups	boiling water	750 mL
2 cups	sliced, fresh strawberries	500 mL
1 tbsp.	sugar	15 mL

Drain pineapple, reserving the liquid. Pour 1 cup (250 mL) hot water over gelatin in a medium sized bowl.

Add cold water to pineapple liquid until it measures 1 cup (250 mL). Stir into pineapple flavoured gelatin. Refrigerate for about 1 hour or until gelatin is thickened to the consistency of egg whites.

Cover the cut up rhubarb with the remaining 2 cups (500 mL) hot water. Let stand for 3-4 minutes, then drain.

In a medium sized bowl, layer rhubarb, pineapple and strawberries starting with rhubarb. Sprinkle sugar evenly between layers of fruit.

Let stand while gelatin is thickening in refrigerator.

Gently fold fruit into gelatin. Spoon into a 6 cup (1500 mL) jelly mold. Refrigerate until firm, at least 2 hours.

To unmold: run a small spatula around edge of mold. Invert over a serving plate onto a bed of sliced lettuce, if desired. Shake gently to release. If necessary, place a warm cloth over inverted mold; shake again to release. Serves 8.

Variations: Substitute 2 cups (500 mL) slightly crush fresh raspberries for strawberries.

Rhubarb Cocktail

4 cups	rhubarb	1000 mL
2 cups	water	500 mL
1½ cups	sugar	375 mL
	lemons	2
	orange	1

Squeeze juice out of lemons and orange. Set aside. Chop and wash rhubarb. Place in large saucepan and add water. Grate peel of lemons and orange. Add the grated rind of the orange and lemons to rhubarb. Cover and simmer.

When rhubarb is soft, strain through a sieve. Dissolve the sugar in the hot water; add to the rhubarb juice. Add the juice of lemons and orange. If a clear juice is desired, strain through cheesecloth.

A tart, refreshing drink. Crisp and cool on its own or with white wine, 7-up or vodka added. Makes 3 cups.

Rhubarb Wine

Our neighbour Bill says of all the wine he's made, this has the prettiest colour.

12 cups	rhubarb cut into 1" (2.5 cm) pieces	3000 mL
6 cups	white granulated sugar	1500 mL
1 tsp.	acid blend	5 mL
1 tsp.	yeast nutrient	5 mL
1 gallon	water	4 L
2	campden tablets	2
¼ tsp.	grape tannin	1.25 mL
1 pkg.	wine yeast (andovin)	1

Cut up rhubarb and put in primary fermentor. Pour dry sugar over fruit to extract juice. Cover with a plastic sheet and allow to stand for 24 hours.

Add all of the other ingredients including wine yeast. Ferment 48 hours.

Strain out pulp and press as dry as possible. In 3 or 4 days, syphon liquid into gallon jugs or carboy and attach fermentation lock. Rack in 3 weeks. Make sure all containers are topped up. Rack again in 3 months. When wine is clear and stable, bottle.

Wine may be sweetened to taste with sugar syrup at time of bottling (2 parts sugar to 1 part water).

Add 3 stabilizer tablets per gallon to prevent renewed fermentation. To preserve colour and flavour, add 1 antioxidant tablet per gallon. Age 6 months.

Rhubarb Punch

This recipe brings back memories of large family barbecues and slow, hot summer evenings.

16 cups	rhubarb	4000 mL
8 cups	water	2000 mL
3 cups	sugar	750 mL
3 cups	boiling water	750 mL
¾ cup	lemon juice	185 mL
1 - 12 oz. tin	frozen orange juice	339 mL
4 qts.	soda water	4000 mL
	mint	

Chop rhubarb and cook in water until tender. Strain through a jelly bag without squeezing. Dissolve sugar in boiling water and add to rhubarb juice.

Chill and add remaining lemon and orange juices. Just before serving, combine with soda and add crushed ice cubes and sprigs of mint. Makes 7 quarts (700 mL) or 70 punch cupfuls.

Rhubarb Concentrate

4 cups	rhubarb, fresh or frozen cut into 1" (2.5 cm) pieces	1000 mL
3 cups	water	750 mL
1	whole allspice	1
7	whole cloves	7
¼ cup	sugar or sugar equivalent	60 mL

Combine rhubarb, water, cloves and allspice in a stainless steel or enamel saucepan. Bring to a boil. Reduce and simmer, uncovered, for 30 minutes.

Add sugar or sweetener. Stir until dissolved. Strain for 3 hours in a jelly bag of several thicknesses of cheesecloth. Discard pulp. Pour juice into a container and refrigerate or store in freezer in small, covered containers.

To serve: mix concentrate with equal amounts of water, ginger ale, soda water or white wine. Sweeten to taste, if desired.

Rhubarb Leaf Aphid Spray

8 cups	water	2 L
3 or 4	large leaves 12-16″ (30-40 cm) across torn up into hand-sized pieces	3 or 4
¼ cup	pure soap flakes	60 mL

Simmer leaves in water until liquid is half boiled away. Pour through a strainer into a bowl or bucket. Mix in soap flakes. Cool. Pour into a capped bottle and store in a cool, secure place until ready to use.

To use: dilute ½ rhubarb leaf mixture and ½ water. Spray on aphid-infested plants.

The New York Tribune reported in 1943 that Mr. "Red" Barber, announcer for the Brooklyn Dodgers, used the term "rhubarb" to describe a row on the field or a mix-up in play, hence the use of the term "rhubarb" today as slang meaning a heated arguement or row.